Adventures of an Alaskan Woman Biologist

by

Margaret F. Merritt, Ph.D.

RDS Publications
Fairbanks, Alaska

Published by
RDS Publications
Fairbanks, Alaska USA
rdspublications@gmail.com

Library of Congress Control Number: 2023900096

Library of Congress Cataloging in Publication Data:
Merritt, Margaret F. 2023
Adventures of an Alaskan Woman Biologist
Includes references and original images

1. Sockeye Salmon at Chilkat Lake 2. Salmon Studies in Juneau 3. Game Management in Palmer 4. Kotzebue 5. Copper River Salmon 6. Lower Cook Inlet Shellfish 7. Statewide Shellfish Research 8. Yukon River Chinook Salmon 9. Seward Peninsula Arctic Grayling 10. Research Supervisor 11. Stock Assessment in Region III 12. New Ways to Solve Old Problems 13. More Land and More Governance

Printed in the United States of America

ISBN 978-0-9828392-6-3

Cover photo was taken by James C. Lund in the Wrangell Mountains. Back photo was taken by Norma Wolf Dudiak in Lower Cook Inlet.

Dedicated to my early biology teachers who created fun biological adventures and inspired me to persevere in my studies

Eugene Andrews at Spring Valley Junior High School
John Burak at Monte Vista High School
Rudy Ruibal at the University of California, Riverside

Contents

Preface

In this book, I share my adventures as one of the early women biologists with the Alaska Department of Fish and Game (ADFG). I portray life as a field biologist working across Alaska's varied landscapes and water bodies, relating observations of the wild animals I encountered and the challenges of studying them. From aerial moose surveys to sonar salmon counts to king crab index fishing, the accounts describe the struggles and triumphs in collecting information needed to ensure that conservation measures sustained the harvest of fish and game for public benefit. It was not lost on me that people's livelihoods or food supply depended on my ability to succeed in my work. I felt great purpose in knowing that my research mattered to the economic and social well-being of the community.

My story spans 1977–2001, a time in Alaska's history shaped by oil wealth fluctuations, the federal takeover of subsistence management on federal public land, and the disastrous *Exxon Valdez* oil spill. I relate my experiences chronologically, partitioned into chapters by job and location. I also offer a glimpse into my personal life as I embraced the pioneer spirit of self-sufficient bush communities by hunting and fishing. As a volunteer emergency medical technician in rural areas, I detail ambulance runs that remain sharp in my memory. And I touch on integrating motherhood into my career.

My tale begins in Southeast Alaska, where I counted salmon at the Chilkat Lake weir. In 1978, I conducted aerial moose counts in Southcentral Alaska before moving to the Northwest Arctic, where I researched chum salmon and caribou. My first permanent position as a biologist came in 1981 when I supervised the enumeration of salmon returning to the Copper River. Subsequent assignments took me to Lower Cook Inlet and Norton Sound to conduct shellfish research. During the turmoil caused by layoffs in 1986, I gratefully landed on the Yukon River Chinook salmon stock biology team. In 1988, I joined the Sport Fish Division to study Arctic grayling on the Seward Peninsula. In 1989, the *Exxon Valdez* oil spill upended many people's plans, and I was called back early from maternity leave to fill in for staff summoned to the site of the catastrophe. For the next twelve years, I supervised sport fish research in Region III, which came to encompass 526,000 square miles, or 89 percent of Alaska's land mass, with waters in the Arctic and Yukon, Kuskokwim, and Upper Copper/Upper Susitna River drainages.

Biologists generally have a special empathy toward the natural world, and that holds true for me. I enjoyed the outdoors, delighted in the animals, and was curious to find answers to biological puzzles. But I discovered greater fulfillment in my interactions with people. I recall visiting with villagers in Noatak over a cup of coffee as they shared their stories about salmon and with fishermen in Homer who described their catch of the day.

My experiences in sampling and observing animals highlight the issues that arose and the problem-solving steps I took, alone or with my team, to achieve the desired goals. The heart of each story lies in overcoming ever-present obstacles. Much of field research involves problem-solving. Most of the time, problems such as gear failure or unexpected animal behavior were irritating but straightforward to address. However, sometimes problems erupted that threatened my safety. Survival depended on my skills, tenacity, and luck. Field research in Alaska is not for the faint of heart, for Alaska's wilderness holds a deadly beauty. To date, twenty-seven ADFG biologists have died in the line of duty. I, along with my colleagues, at one time or another, have wondered whether we would join those somber ranks.

I was the first woman to undertake the job in many of my assignments, which sometimes caused the men I worked with to react with defiance. Some resisted the inclusion of women in the workforce because it challenged their preconceived ideas of what a stable workplace looks like. My response was to act as professionally as possible, even when instances called for me to confront misbehavior. Thankfully, I wasn't always on my own. Throughout my career, supportive supervisors and colleagues took a chance on me. They provided the opportunity, and it was up to me to prove my worth.

Knowing I was the first woman to tackle a job was sometimes isolating. I felt pressure to perform excellently, as I believed how I was perceived might affect future opportunities for women. To enhance my credibility, I completed a Ph.D. degree in Fisheries. I was able to contribute increasingly innovative and technical advice to staff discussions and so gained credence. Did my job require a Ph.D.? No, but it gave me the standing to navigate as a woman in a male-dominated profession.

The book is a tale of growth for me and ADFG. I learned the expertise and attitudes necessary to work and survive in challenging field situations and how to apply problem-solving tools. My confidence grew with each harrowing escape from disaster and each hard-won professional success. I watched as ADFG biologists argued with and navigated the inevitable changes that advanced upon them: women entering their ranks, biometricians inspecting their work, and federal forces directing changes in land conservation and subsistence regulations. Reluctance to change occurs today in many parts of society, but the strong dedication that ADFG biologists have toward their mission encourages me to think they're prepared to incorporate change when needed.

The characters in the book are denoted by their title, an alias, or their name. Source material included my logbooks, letters, and publications. Reference numbers are unique and are in the order in which they first appear, with subsequent citations retaining their unique numbers. The author provided all unsourced images.

Acknowledgments

I greatly appreciate the insightful review comments offered by Laura Lund, Cathy Lair Klinesteker, Elise Israel, and Joe Shaw. Norma Wolf Dudiak, Nick Dudiak, Jim Lund, and Jack Whitman kindly allowed the presentation of their photographs in this book. My thanks to Cynthia Frank at Cypress House for her enduring enthusiasm and helpfulness. Patricia Rasch added creative polish to the maps and cover design.

Prologue

In June 1977, I submitted my thesis in partial fulfillment for my Master's degree in Biology-Ecology at Utah State University. Just the defense was left, scheduled for the end of August. For three years, I studied in preparation for a job as a biologist, focusing on the adaptations of small desert mammals to their habitat.

At the start of my graduate studies, I obtained a loan to buy a small house in Logan, complete with a garden, and upon graduation, I hoped to find a biologist position in the area. However, job inquiry letters mailed to agencies and environmental consulting firms came back negative.

I had some free time on my hands before my defense. A friend of mine had just moved with his family to Sitka, Alaska for a job as a floatplane pilot at an air taxi service, and he sent me a postcard saying I should come see the country. I had thought a lot about visiting Alaska and was eager to go. My dad had traveled to Alaska in 1956, and I remembered his stories of the country he had seen. At age twenty-five, I set out for Alaska.

I wrote to my mom about my travel plans, "I'll sightsee and look for jobs while I'm there. I'll be in Alaska for only two weeks, I guess." After arriving in Bellingham, Washington, I spent $85 on a ferry ticket and walked onto the Alaska Marine Highway ferry for Sitka.

Sockeye Salmon at Chilkat Lake

"A lake is a landscape's most beautiful and expressive feature. It is earth's eye; looking into which the beholder measures the depths of his own nature."
~Henry David Thoreau

The ferry nosed into the Sitka harbor on a brilliantly sunny day in July. I inhaled the pungent scent of seaweed, salt splash, and diesel from nearby commercial fishing boats. My senses felt enlivened as I disembarked with my small backpack, stepped on solid ground, and walked into town. I discovered that Sitka, located on Baranof Island, is a small city rich in culture and history. The Tlingit people first settled on the island thousands of years ago. In the early 1800s, the Russian American Trading Company gained control of the site to strengthen its trade routes on the Pacific West Coast. In 1867, Sitka was the site of the signing of the Alaska Purchase when Secretary of State William Seward bought Alaska from the Russians for $7.2 million.[1]

My friend was working, so I set off to explore the nearby park and the Sitka Museum. The next morning, he took me flying in a Cessna 185 on floats. Skimming the water and then lifting off was a new and fun experience. We flew above Baranof Island, an emerald gem of dark green forests set against a sparkling blue ocean. Landing on floats was even more fun, bouncing on small waves and timing the glide just right to the dock.

Full of enthusiasm after my morning sightseeing flight, I grabbed my envelope of résumés and set out to find the Alaska Department of Fish and Game (ADFG) office in Sitka. The old wooden building was small. It had a homey atmosphere, warmed by the woman at the front desk, who greeted me with a smile. I explained that I had just arrived in town and asked if there were any job openings. She brought out two men from back offices, one from the Game Division and one from the Commercial Fisheries Division.

We all sat at a small conference table where I handed out my résumé and answered their questions about my education and experiences. Within the hour of my stepping into that old wooden building, I was offered two jobs, starting immediately. The jobs were temporary, which meant they were short in duration, and I would not get any benefits, but hey, they were job offers!

The tough part was deciding which job offer to accept. I didn't even question my instinct to take one. The Game Division's job to map wildlife habitats lasted one month and was based in Sitka, so I would have to find housing and pay for food. The Commercial Fisheries' job to count salmon lasted two months and was located at a lake near Haines, at the top of Lynn Canal, with transportation, room, and board paid for by the state. I told them I needed a moment to think about it, so I walked around the block while scenarios and uncertainties raced through my mind. Twenty minutes later, I accepted the job with Commercial Fisheries because it would pay more, and I would see new country for free. But my acceptance came with a caveat. I needed a week at the end of August to fly back to Utah to defend my thesis. With a nod of agreement, I was told I would begin work on August 1, just three days away. Arrangements were made to leave tickets for me at the airline counters for my flight from Sitka to Juneau, with a connecting flight to Haines.

Funny how decisions made in your twenties, during that age of exploration and seemingly endless possibilities, can set you on a path for a lifetime. I'm tempted to think about that twenty-minute walk I took around the block so many years ago in Sitka and wonder where my path would have led if I had made a different decision.

But what is important is to fully live the life I have, not to imagine what might have been—an impossible task, anyway.

I was unaware of the impact that the construction of the Trans-Alaska Pipeline (1974–1977) had on Alaska. In the 1970s, state jobs were hard to fill because everyone wanted to earn the high wages offered by the oil companies. Time and again, the state hired and trained workers, only to have them leave for a pipeline job. The state began to save time and money by creating temporary positions; hiring temporary employees required less paperwork, and they were not paid benefits. Temporary workers were also not required to pay the Alaska Public Employees Union dues. Over time, this thorny issue prompted labor negotiations that greatly impacted my life. But in July 1977, I was oblivious to background currents and blessed my good fortune at finding a job fresh out of college.

I told my friend that I had found work and would be leaving Sitka. He saw how excited I was to have a job as a biologist, so he wished me good luck and bid me farewell. I was so busy I didn't have time to drop a note to my parents until I arrived in Haines, waiting to meet my new boss. "Dear Mom," I hurriedly wrote, "The day after I got to Sitka, I went into the ADFG office with my job portfolio and was offered two temporary jobs. I took the one at Chilkat Lake by Haines. The job pays $1,214 per month, with room, board, and transportation paid for by the state. The sockeye salmon are running, and I'll work on their census. I'll live with another girl in a cabin on the lake. I start tomorrow and will work three weeks, then have a week off to defend my thesis in Utah. I'll go back to the lake for another month until the end of September. I hope I'll be offered a permanent job."

Haines sits on a sliver of land at the head of Lynn Canal, between Chilkoot and Chilkat Inlets. The Tlingit word "Chilkat," translated into English, means "salmon storehouse." Sockeye salmon destined for Chilkat Lake swim up Lynn Canal, Chilkat Inlet, Chilkat River, a slough, and then arrive at the lake where they spawn. The harvest of sockeye salmon was the primary reason for early human settlement in the area.

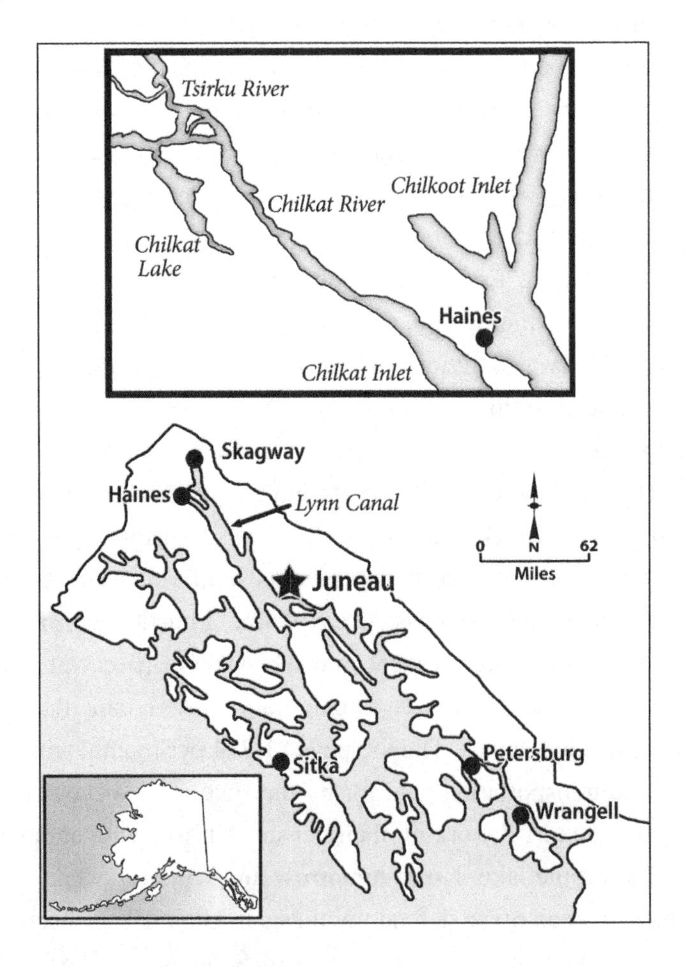

My new boss, John, found me at the airport, and we drove to where the ADFG boat was tied up.

"Do you have hip boots?" John asked as he transferred boxes of groceries into the boat.

Upon receiving my negative reply, he rummaged in the back of the truck until he found two boots and handed them to me, my first issue of state property. John wasn't much for words, but he filled me in about the history of the commercial fishery and my job. Commercial fishing in Lynn Canal began in the late 1800s. Early commercial harvests averaged over one million fish annually. During the 1970s, the average annual drift gillnet commercial harvest of sockeye salmon in Lynn Canal was around 122,800.[2] Commercial harvest of sockeye salmon is regulated through time, area, and gear restrictions to achieve an escapement range (the number of salmon counted on their spawning grounds).

The fish mature at sea between ages three and seven and return to their natal freshwater lake to spawn. To get to their spawning grounds in Chilkat Lake, the salmon pass through a slough from early June to late September. In 1967, ADFG erected a steel picket weir across the slough to count their passage. Technicians open a gate at regular intervals to count the fish as they swim through. Daily counts of sockeye salmon passing into the lake are relayed by radio to the office. If counts are lower than the historical average, managers reduce harvest. If counts are on track with the historical average or higher, that suggests a good run, and managers consider allowing more harvest. My work would help to determine the income of commercial fishers.

Because I had a graduate degree in biology, my rank as a Fishery Technician III meant that I would supervise the other girl at the lake, Amy, who was a Fishery Technician II. Amy was born and raised in Southeast Alaska. She had experience operating weirs and had begun work at Chilkat Lake in June. Apparently, someone before me had quit, and I was the replacement. John introduced me to Amy, explained that I was in charge, told Amy she was to show me how things were done, unloaded the boxes, and took off. As the sound of the outboard motor died away, it was just the two of us standing in a small cabin in the middle of the wilderness. Amy let me know she disapproved of a "Cheechako" (an old gold rush term for an ignorant newcomer to Alaska) as her boss. After all, she knew more about it than I did. I was determined to get along, so I let her grumble as I unpacked and looked around.

ADFG cabin at Chilkat Lake weir, 1977

The inside cabin walls were unfinished, with pink fiberglass exposed and hanging at precarious angles. I'd have to order plywood to be brought up with the next load of groceries to finish the walls. My bunk was on top. The radio was in the kitchen. The outhouse was located a good distance away and accessed by a walk through the marsh. A wood plank walkway lay across the weir. The weir's location was perfect: the water velocity was slow to moderate, water depth was 4 to 5 feet, there were no undercut banks, and the stream bottom of coarse gravel was level and uniform. I noticed that the banks were not very high; in fact, the whole area was flat.

I watched Amy as she carefully made her way across the planks to the middle where the gate was located and pulled out stakes. The stakes were inserted through holes in steel panels and set at an angle. Sockeye salmon that had amassed on the downriver side of the weir eagerly swam through the opening. Amy pointed out that sockeye salmon are sometimes called "red" salmon because their bodies become red during spawning while their heads turn green.

Weir across the slough leading to Chilkat Lake

Amy used a mechanical counter to record fish passage. After letting the salmon pass, she replaced the stakes and ensured the ends were firmly anchored into the stream bottom so no fish could swim underneath. We picked off debris that washed against the upriver side of the weir, a twice-daily routine to prevent a large debris load from washing out the weir.

On my second day in camp, Amy slyly told me it was my turn to take the garbage out to burn. She handed me a gas can that had "Blazo" printed on the side and said to pour it onto the trash after I had dumped it in the pit. Then, set the garbage alight. Never having heard of Blazo before, I doused the contents over the garbage and dropped in a match. Kaboom! A big explosion hurled me on my butt 4 feet away and scorched my eyebrows. She was chuckling when I returned to the cabin, mad as hell. I could have been hurt! Feeling a bit contrite, she explained that Blazo, an Alaskan Bush staple, is highly flammable and perfect for starting a fire if wood is wet, as it often is in Southeast Alaska.

Amy and I edged into a truce, and I settled into the daily routine of weir life: count fish, radio in results, clean weir. In my time off, I paddled the canoe up the slough into Chilkat Lake, nestled at the base of the snow-capped Chilkat Mountains. As the canoe nosed into the lake, I looked down to watch salmon below the surface. Females hovered over depressions called "redds" they had made in the gravel with their frayed tails, spilling eggs. Bright red males darted around females, eager to expel their sperm in the fertilization ritual. Other salmon swam lazily, their energy spent and their life coming to its close. Salmon can live up to several weeks after spawning before their carcasses begin to decompose.

A salmon spawning ground has a rotten fish odor that is attractive to scavengers. On my first trip to the lake, I expected to see seagulls and maybe glimpse bear and fox tracks. As my gaze finally shifted from below-surface activities to the skyline, I was surprised to see the moss-draped spruce trees at the lake's edge heavily laden with hundreds of bald eagles. The birds had perched after gorging on decaying salmon. I froze in the canoe, fascinated by magnificent white heads illuminated against the green foliage, turned toward me in attention, their gaze calm and unrelenting. I later learned that Chilkat Valley is home to the world's largest concentration of bald eagles. In the fall, over 3,500 bald eagles descend on the area to tear into salmon carcasses. In the late summer of 1977, I had the lake and the eagles all to myself. Five years later, in 1982, the valley was designated an Alaska Bald Eagle State Park and Preserve.

I beached the canoe on a sandy bank and settled in the sun on a log to watch other lake inhabitants. Kingfishers dove into the lake to catch fish. A furtive red fox padded to a salmon carcass washed ashore for a few quick bites before dashing into the underbrush. Harlequin ducks ruffled themselves in the shallows. A magnificent way to spend my free time! I was happy there in the wilderness. As I sat still, tuning my senses to the buzzing sounds and movements around me, I became aware of the artistry of Chilkat Lake. Life at the lake was interrelated in an ingenious and colorful mosaic, and I felt part of it.

Some days, white mist hung low in pockets at the lake's dark edge, casting the lake in grayscale. The lake's surface was glass on these calm days, and the forest was hushed. On other days, the bright sun released a palette of colors above and below the lake's surface, which shifted and danced in rhythm with wind-driven ripples. Energized by sunlight, the lake's aquatic inhabitants darted and splashed in their hunt for food while avian creatures chatted invitations or called warnings.

The health and diversity of life at the lake depend on salmon. Their bodies contribute marine-derived nutrients to plants and animals living in the riparian community through direct consumption by scavengers, who drag disintegrating carcasses onto land, and through decomposition. Critical nutrients of life, such as nitrogen, carbon, phosphorus, and calcium, leach from salmon carcasses into the watershed and are taken in by algae and protozoans, food for copepods, who are in turn eaten by juvenile salmon, in an intricate feedback loop.

I've had an interest in nature since I was a little girl. My dad took me with him on his trips to the desert to photograph cacti in bloom, and we'd roam the hills looking for geodes for our rock collection. My mom loved cabin retreats in the mountains, where she'd set up an easel to paint while I settled under a tree to read. Such books as the *Leatherstocking Tales* kindled my interest in exploring the wilderness, while *Walden* offered more detailed and thoughtful ideas about the living world.

Our communication with civilization was the radio that sat in the kitchen. Amy delighted in ordering an amazing variety of food items over the radio. "It's an unlimited food budget," she explained. "We can order whatever we want." Once a week, John motored up with a delivery of mail and groceries that included seafood, steak, fancy desserts, and imported Norwegian goodies—additional compensation for working in a remote camp.

In addition to reporting daily counts of salmon that passed through the weir, we also reported other activities, such as boat traffic headed to the lake. Few boats came to the lake in summer, but traffic increased once duck hunting season opened. We could hear an outboard motor approach long before it reached the weir, so we were ready when it arrived. We were obligated to allow boats to pass upstream, so we pulled stakes at the gate. Boat traffic disrupted salmon counts because salmon could escape upstream in the boat's wake unseen and, more importantly, uncounted. Thus, a report of boat traffic suggested counts were a minimal estimate for that day.

A few weeks after my arrival, we began to hear, late at night, the sound of an outboard motor. We thought it was odd that a boat never appeared at the weir to ask for passage; else, why would they be on the slough? We reported the erratic late-night sounds.

On John's next delivery of mail and groceries, he announced news of a covert operation that Fish and Wildlife Protection, an arm of the Alaska State Troopers, was setting up downstream of the weir. Those late-night motor sounds we heard were poachers illegally netting salmon. Salmon will mill downstream of a weir until they find passage, thus making them vulnerable to easy netting. Regulations prohibit fishing within 300 feet of a weir. John told us to lock the cabin door that night and not come outside, no matter what. In the night, again, we heard an outboard motor, then faint shouts and gunshots! We didn't sleep much the rest of the night, wondering what had happened downstream. The next morning, on the radio, John explained that the enforcement operation had succeeded. The culprits had been caught in the act of illegally netting salmon. Those shots we heard were just fired as a warning. No one was hurt. We appreciated quiet nights for the rest of the season.

The last week in August, I stepped into John's boat with my pack, and we headed downriver. I was on my way to Utah to defend my thesis. Amy would have John for company during my absence. As I walked along the dock in Haines, it felt good to see civilization again. My trip was a flurry of activity between my house business and school business. After my thesis presentation, I met with committee members in a small conference room. They took turns asking questions and offering comments. I passed. Hooray! With minor corrections completed, I turned in my thesis to the Graduate Studies Office. I was done with school, and I felt free. I still had one more month of work left with ADFG.

Amy was actually happy to see me when John brought me back to the weir in early September. We easily settled into our old routine. In my off hours, I canoed to the lake and read a stack of new books. I also brought along reports to finish. While in school, I worked as a biologist for the Desert Biome, part of the U.S. International Biological Program funded by the National Science Foundation that aimed to study and model ecosystem productivity. At the Curlew Valley Validation Site in the Great Basin Desert, I was in charge of small mammal population studies and monitoring soil-seed-biomass reserves. While I enjoyed the research and benefited immensely from the field experience, the project was ending. I was eager to close out that chapter of work and focus on future research possibilities.

With the advance of fall came bursting rainstorms. The slough flowed full and swift, whirling dark debris mats of leaves and wood that we cleaned off the weir throughout the day. At times, big tree branches and even entire trunks were forced up against the weir, and we had to saw off pieces to wrestle either over or around the stakes. We set up nightly weir checks with flashlights to monitor the debris load. Early one morning, we awoke to a wet cabin floor. The entire site was flooded! The slough flowed over its shallow banks and moved thickly through submerged bushes while it generated fat yellow bubbles of scum in the eddies. The outhouse was afloat, and salmon freely darted past my hip boots, the green-headed devils gleeful in their high-water dash to breed. What a mess! We immediately radioed John that the weir was out of operation; there was nothing we could do about the situation until the water level subsided.

Winds scuttled rain clouds out of the valley, and three days later, the slough flowed through new banks, carved in a patchwork pattern through brush and grass. Amy and I worked hard to cut off and haul away debris lodged against the weir and reset bent or uprooted stakes. John brought heavy steel panels and stakes that we erected at the edges to direct migrating salmon toward the gate. The weir washout allowed hundreds of salmon to pass upstream without being counted, inserting a degree of uncertainty in managing the commercial fishery during those three days.

Salmon counts resumed, as did the progress of fall. The air became colder, and on September 21, 1977, I awoke to new snow on the mountains surrounding Chilkat Valley. Salmon counts markedly decreased; the sockeye salmon run was nearing its end. Plans were made to dismantle the weir for the season, and my first job with ADFG came to a close. In 1977, the Chilkat weir operated from June 3 through September 27. A total of 41,044 sockeye salmon were counted as they passed through the weir.

In my time at Chilkat Lake weir, I learned to live in the salmon's world. I monitored their migratory behavior, witnessed their breeding rituals, and observed their beneficial relationships with the lake's inhabitants. As stewards of salmon, ADFG negotiates a balance between human harvest and ecological sustainability.

Salmon Studies in Juneau

"We can't become what we want by remaining what we are." ~Max DePree

I forwarded my Haines, Alaska, mail to San Diego, California. In early October 1977, I spent much-needed time at home with my father. I was shocked to see how aged and weak he had become. I also went to the nursing home to visit my cousin Emma. I fretted about leaving my elderly family, about their care and companionship. And mother, I worried about her in Oregon. While Mom was fearless about going forth on her own, the toll of hard times sometimes weighed her down.

By October 18, I was back at my house in Logan, doing maintenance and harvesting the garden. I was relieved that my roommates had taken good care of the house while I was in Alaska. But I had little peace of mind with the gnawing need for an income. I began talking to people and searching for connections to create another job opportunity.

It came about that a friend, Sue, was dating the brother of a man who worked at ADFG in Juneau, and after a few phone calls, by October 20, I was offered a position in Juneau starting November 1. The job was a temporary Technician III with Sport Fish Division. It was only a month, but I felt confident that doors at ADFG would open for me if I kept knocking, so I accepted the job. I told my roommates I was leaving for Alaska again, called my parents, took care of last-minute business, and booked a flight out of Salt Lake City. The twenty-fifth year of my life had its share of uncertainty and adjustments, but it was also a time of possibilities.

Juneau, the charming capital of Alaska, is located on a strip of temperate rainforest wedged between mountains holding back glaciers and the Gastineau Channel. The city was named for Joe Juneau, who in 1880 found gold on a mountain in the channel. The discovery of gold transformed the site into a thriving mining and supply center. Diverse groups of fishermen, legislators, natives, tourists, and an occasional lumberjack closely mingle in the narrow streets and small shops and give the city an unpolished cosmopolitan air. No roads lead to Juneau; you arrive by boat or air.

I flew into Juneau with a cooler of frozen venison and 20 quarts of fruit and vegetables from my garden. I found a small hotel room for $18 a night. The next day I looked up a friend of a friend to say "hi" and get acquainted. He lived with other young professionals, and they were looking for a roommate. I was overjoyed as housing was scarce. I moved into a big, old drafty house with four bedrooms on a steep hill overlooking Juneau. The view of the Gastineau Channel from the living room windows was beautiful.

View of Gastineau Channel from our house in Juneau

There were six of us: a planner, two lawyers, a nurse, a clerk, and me. The house's owner was returning in January, so we would all have to move in a few months, but it was good housing for the time being. My new roommates were helpful and fun; we laughed a lot. Cheryl, the clerk, took me on a tour of Juneau. Several of us chartered a plane to view the Juneau Icefield. Liz, one of the lawyers, became a friend and, with borrowed gear, took me snowshoeing and cross-country skiing. Matt, the planner, showed me hiking trails and nearby mountains.

I discovered that winter in Juneau means cold rain, wet snow, and roaring wind. Shortly after I arrived, a Taku windstorm came. Air turbulence spilled from the mountains through Taku Inlet and hit the city with hurricane force. I fought against the cold wind to get home after work, grabbing onto lampposts to keep from being blown down as I slowly worked my way up the hill. The wind tore at my clothes and pelted my eyes. I was exhausted and strangely exhilarated to finally step into the quiet of my home. I was happy and optimistic despite only seeing the sun once in two weeks.

My job with the Sport Fish Division was to prepare and "read" salmon scales for age determination. My boss, Paul, explained that salmon scales form ridges (circuli) that radiate from the center (called the focus), thereby recording the life history conditions of the fish, similar to rings on a tree. Rings of narrowly spaced circuli (called an annulus) indicate slow growth in winter when food is scarce and the water is cold. Rings of widely spaced circuli indicate rapid growth in summer when food is abundant and water temperatures are warmer. The pattern of growth rings determines age: each winter annulus represents another year of life.

Technicians take scales from a sample of salmon harvested throughout the fishing season at boat landings or angler sites. Spawning salmon are sampled in streams. Fishery managers use age information to estimate how many fish of a given brood year return to spawn (stock productivity), forecast future fish returns, and adjust harvest strategies to ensure the desired number of salmon reach their spawning streams.

Salmon scales are always taken from the same "preferred" area: on the left side of the fish, two rows above the lateral line along a downward diagonal from the dorsal fin to the anal fin. Two to three scales are plucked from this area using forceps, cleaned of slime and dirt, moistened (do not lick), and placed on a yellow gum card, with scales all oriented in the same manner. Scale ridges are faced outward—you can feel them with your fingertip.

Information for each fish (species, date, location, sex, and length) is recorded on the card. Each gum card can hold the scales from thirty fish. In the field, gum cards are stored between sheets of wax paper so they don't stick together.

Paul impressed me with stories of collecting scale samples from salmon in streams frequented by brown bears. With a graphic illustration of the seriousness of biological fieldwork in Alaska, he rolled up his pant leg to reveal a long red gash. While walking along a small stream thickly overgrown with vegetation looking for salmon, a bear silently reached out of the bushes and pulled him down. Paul had no idea he was within a few feet of the hidden bear. His body was pinned, suffocated by the bear's mass of rippling muscle and hair. He felt faint from the bear's grunting, warm breath as it fell, moist and foul on his face. Paul freed up an arm to unholster a pistol strapped to his hip and pulled the trigger, firing a bullet wildly into the air. The noise startled the bear, who ran off, saving Paul from further mauling or death. Paul's strong advice was to always carry a pistol when sampling salmon in streams.

While a pistol is no match for a brown bear, the problem is that the shotgun issued by the state for bear protection always got left in the boat or propped up on a rock and was not at hand when you needed it. Paul's advice stayed with me for the rest of my career. In years to come, when I walked streams for salmon carcasses, sure enough, we always left the shotgun in the boat or lying on a rock. How can you carry a shotgun when you've got your hands full of decaying salmon flesh, forceps, and gum cards? Like Paul, I had a pistol strapped to my hip: a six-shot .357 magnum with a six-inch barrel. I never had to use it, but it was reassuring.

I was assigned to a small office with a tiny desk on which were piled hundreds of yellow gum cards. I organized the stacks according to Chinook, sockeye, or coho salmon and then carried boxes of gum cards to the lab to make impressions in clear acetate with a hydraulic press. The right amount of heat, pressure, and time had to be used to make impressions, or the scale ridges would melt together. On occasion, I encountered errors: scales were mounted the wrong side up so that no age could be determined, or annuli were difficult to see from slime and dirt left on the scale. Some gum cards were stuck together.

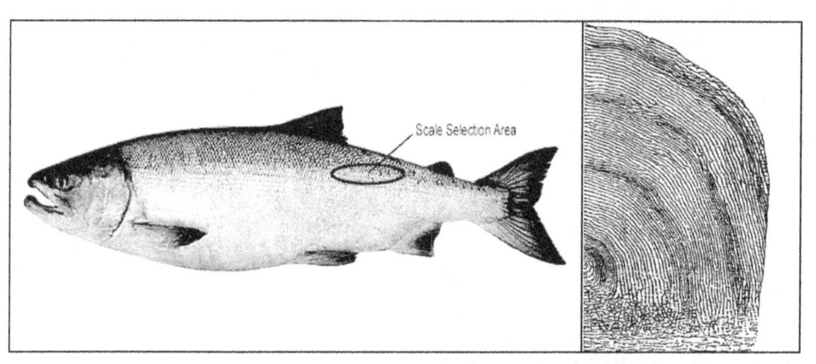

Preferred area where scales are sampled from salmon for age determination

Scale from an age 1.3 salmon

I "read" scale ages by inserting acetate cards into a microfiche that magnified the scale image. To determine age, you start at the focus and count the number of annuli, progressing toward the edge of the scale. The predominant age class for sockeye was 1.3 (where

"1" denotes the number of years the fish spent rearing in freshwater, and "3" is the years the fish spent feeding in the ocean before migrating back to its home waters). Counting one year in freshwater and three years at sea, the returning salmon was four years old. After hours of reading scales, my eyes needed a rest!

To break the monotony of scale reading, I asked for access to a computer so I could set up files and analyze age data. I was curious about what the scale age data revealed. And I knew bosses value people who look around and do work that needs to be done. For me to get noticed by my superiors and to become a biologist, I needed to expand my work beyond the sphere of a technician's duties. I had the desire to do more, but importantly, I could do more because of my education. It takes more than a willingness to reach for higher goals; it also takes preparedness.

I wrote to Mom, "I haven't been paid yet and have just $15 left. I'll have to wire for money from Logan. I'm applying for other jobs with ADFG and had an interview this week."

My familiarity with computers and data analysis was noticed around the office. At the end of November, I was notified that I had been hired as a temporary Fishery Biologist II with the Fisheries Rehabilitation, Enhancement, and Development Division (FRED), created by the Alaska Legislature in 1971 to produce salmon in areas where they were in decline and creating economic hardship. The job ran from December 1 through April. Although temporary, it would get me through the winter.

In 1977, FRED was in a growth phase and building hatcheries. At first, everything was disorganized. I had several bosses who told me different things to do. I found that one boss, Karen, was sensible, so I began to clear assignments through her. Work began to flow more smoothly.

One task was to design experiments testing the effectiveness of various incubation strategies on salmon egg development and survival at Snettisham Hatchery, located south of Juneau. I was flown to the hatchery on a chartered floatplane and given a tour by the hatchery manager, who explained his concerns. Variables to test in the experiments included the type of substrate to use when loading eggs into the incubation chambers, water temperature, pH, oxygen, and egg density. With a new appreciation of the issues and implications of results, I boarded the Cessna 185 for our return trip to Juneau. Within a few minutes, we flew into low clouds with rain. It was so foggy I couldn't see the wingtips. The pilot dropped to the water to avoid flying into one of the coastal mountains, skimming the floats just above the waves. Luckily, we didn't run into a fishing boat and docked the floatplane in one piece. This was my first of many flights when the Alaskan weather channeled my adrenalin and prayers.

My roommate, Liz, became quite ill, and I caught her cold, which turned into bronchitis in that drafty old house. I was sick in bed for over a week. I couldn't go to work. I tried to go once, and they just sent me home after taking one look at my pale face. Here I was, trying to make a good impression at my new job, and I was sick! My parents mailed me boxes of vitamins, herbal tea, fruit, and other food that the mailman left on the porch, which was good since I was too sick to go grocery shopping and all my roommates had left for the holidays. I spent that Christmas sick and alone in a drafty old house. When I finally recovered, my bosses let me make up for lost time by working extra hours; as a temporary employee, I didn't have any paid sick leave.

January 1978 came, and I had to look for new housing. I found a tiny attic apartment on Basin Road, high on the hill overlooking downtown. I subleased the apartment from a girl who would return in April. Basin Road led to the Perseverance and Granite Basin trails, where I often hiked. A fuel oil stove heated the room, water, and my food.

As spring approached, my duties shifted from biometrics to procurer. The new hatcheries were coming online and needed to be stocked. I was assigned to a four-member team to itemize and purchase everything a hatchery would need to operate, a huge task! We made countless phone calls to managers of hatcheries that were already operating to ask for their suggestions. We pored over stacks of equipment catalogs and bought hundreds of items, ranging from office chairs to Erlenmeyer flasks. We had to figure out if items could be shipped directly from the manufacturer to a remote Alaskan location or if we needed to have deliveries made to the office and then flown on a chartered plane to the hatchery. Piles of boxes arrived at the office, and storage became a problem. It was a hectic yet fun time, and I was fortunate to have been hired when the state of Alaska had money.

I was proud to play a small role in Alaska's programs to increase salmon abundance and enhance fisheries while protecting wild stocks. In the 1970s, we thought hatcheries could become a vital solution to problems in salmon management. But I shared concerns that others in the scientific community had about whether the programs worked as intended or did more harm than good in the marine ecosystem. Several things can go wrong in the quest to enhance marine production of salmon, such as the failure to produce fish that successfully recruit to the fishery, the overexploitation of wild fish in the process of harvesting hatchery-produced fish, and competition for food between wild and hatchery juveniles. After all, the goal was to supplement natural production, not to replace or displace it. Subsequent studies of wild-hatchery salmon interactions helped to understand those relationships and improve hatchery programs. For several fisheries, the decision to enhance salmon production came down to the question of available funds and net benefits.

Southeast Life

My new friends were keen to explore trails and mountains in the vicinity and were adamant that I get outfitted with the right gear. They shepherded me through the local shops, and I spent most of my first paycheck from FRED. I felt justified in the extravagance because I had a good job. Wool was the basic material to keep you warm in a winter rainforest, so I bought Woolrich pants, a thick Icelandic wool sweater, a wool hat and balaclava, and wool mittens. Accessories included gators to keep snow from getting in my boots, a bright orange ice ax, crampons, and more.

One of my first Alaskan adventures was learning to ice climb on Mendenhall Glacier, a 13-mile-long extension of the Juneau Icefield that protrudes into Mendenhall Valley, about 12 miles from downtown Juneau. From the Visitor's Center parking lot, we hiked on a trail in a softly falling mist through the rainforest that gradually became steep rock draped with ice fingers. After a 4-mile hike, we reached the edge of the glacier. We roped up. I buckled my crampons onto my boots, took a firm grip of my ice ax, and stepped onto the ice.

For 2 miles, we picked our way across jumbled ice blocks, tested snow patches for hidden crevasses, and scrambled up ice ledges. We avoided seracs, deep cracks in the ice that might topple. We found a stream of water flowing across the glacier's surface a short distance before cascading into a dark blue cave. At the end of the day, I was exhausted and exhilarated by my first ice adventure.

Matt wanted to see Tracy and Endicott Arms, deep fjords located 45 miles south of Juneau, and invited me on the trip. On a sunny weekend, we chartered a floatplane to take three of us and our gear, including an inflatable raft. The pilot deposited us on a rocky beach, and we gave him a pickup time for Sunday afternoon.

We assembled the raft, threw in gear, and motored up Tracy Arm, encountering icebergs. Waterfalls tumbled from towering cliffs edged with Sitka spruce and western hemlock. Matt wanted to see Ford's Terror, a steep and narrow fjord jutting off Endicott Arm.

Ford's Terror at rising tide

As the tide rises, the sea rushes into the narrow entrance creating turbulent currents that can "trap" unsuspecting boaters in the fjord until the tide becomes slack. The tides were changing as we approached the phenomenon. We were thrilled to watch as aquamarine water shape-shifted into white frothing waves but jetted the raft out of there before the currents became too strong to escape. At high tide, we found a good camping spot, tied the raft to a tree and pitched the tents.

The following day was sunny, and we found the raft perched high on a rocky cliff studded with sharp barnacles at low tide. Boy, were we stupid to have forgotten the 16-foot tide-height difference! While moving the raft down to the water's edge, the boys dragged the overinflated tubes across a patch of sharp-edged barnacles. Two tubes ripped, and the raft deflated. Because there wasn't enough patch material for the long rips, we became stranded. We didn't arrive at our pickup location at the appointed time, so the pilot radioed for a search and rescue. The next day, a helicopter found us and landed on the rocky beach; we sheepishly climbed aboard. Matt had to buy more patch material and charter another floatplane to retrieve his raft. Nonetheless, I thought it was a great adventure, full of beautiful scenery, excitement, and lessons learned.

After four solid weeks of rain, the sun came out in early April, and the ADFG Commissioner gave local employees the afternoon off as a "Sun Day!" I was thrilled to see the sun again. That weekend, Matt and I chartered a plane to explore Gustavus, a small community at the mouth of Glacier Bay National Monument, later designated a national park. The Tlingit people had used the area for salmon fishing and berry picking for hundreds of years. Settlers homesteaded there in 1917. I enjoyed watching Canada geese build nests in the dry grass near our camp. The trip to Gustavus was a bittersweet conclusion to my winter's stay in Juneau. With the onset of spring, my job ended, my apartment sublease was up, and it was time to go home.

***

Brother Bill met me at the San Diego airport with bad news—Dad had been diagnosed with cancer and did not have long to live. When we got to the house, I found Dad thin and weak. The time left was so short and precious! We set up a bed in the living room so he could see out the window to his garden. Dad deteriorated rapidly. We had to feed him as he was too weak to hold a spoon. I was washing sheets like crazy until a nurse inserted a catheter. Friends and family called or came by. He recognized people but had difficulty forming sentences and blanked out often. Two short weeks later, just after my twenty-sixth birthday, he slipped away, drawing a last shuddered breath as I held his hand. I don't remember much, except crying a lot. I must have been in shock.

I became tired of packing, dealing with bills, and business while grieving my father's death. Somehow, the house got packed up, and I put boxes and furniture into storage in San Diego. I'd deal with them later. I walked away from my childhood home. I kept Dad's mailing address to receive his remaining bills and correspondence. After Dad's memorial service, I went to see cousin Emma in the nursing home; she had had another stroke. While her hands were not strong, her smile was. I gave her a long hug. I thought of my friends: Susie had two children already, Cathy was expecting her first, and Mary and Elise were newly married. The circle of birth and death was never as vivid as in May 1978.

I had expenses and responsibilities in Utah. My car and house were there, so that's where I went next. Car insurance, new tires, taxes, mortgage payment, and house repairs all added up to $2,000. There were no career opportunities for me in Utah. I decided my best chances of finding work as a biologist were in Alaska.

With mixed emotions, I decided to sell my house in Logan. I had just lost my childhood home; should I get rid of the only other home I had? I reasoned that a house with no job was not a place I could afford to keep for long. I put a For Sale sign up, and within a few days, a young couple made me an offer. With the help of a lawyer, the house was sold. More boxes and furniture were put into storage. I felt very unsettled! It was a hard time. Dad had just died.

My belongings were spread across two states. I had no permanent address and no job. Well, this was no time to feel sorry for myself. I needed to be strong and set about earning a living. With my Ford Bronco loaded up, I waved goodbye to my life in Logan and drove to Seattle to put the car on the ferry.

As I drove off the ferry at the end of July, it felt good to be back in Juneau. However, my plans to look for work with ADFG in Juneau were upended when my friend, Matt, announced he had a new job and was moving to Anchorage in August. Would I like to go with him? We could drive the Alaska-Canada Highway and sightsee along the way. Could I withstand more change and uncertainty? I took some time to think about it and supposed that the prospects of finding work in the larger ADFG office in Anchorage would be better than in Juneau. Plus, I would see new country. I told him, "Okay, yes, I'll go."

Mt. St. Elias from Yakutat Bay

Before we left Southeast Alaska, we filled weekends with adventures. One trip was to Disenchantment Bay, a narrow 10-mile reach of water into which great chunks of ice avalanched from Hubbard and Turner glaciers. We flew to Yakutat with the Avon rubber raft and, after assembling the gear, motored through Yakutat Bay to a cabin rented for the weekend. The spectacular profile of Mount Saint Elias loomed large across the bay.

An iceberg floats in Disenchantment Bay

Disenchantment Bay edges the southern boundary of what became the Wrangell-Saint Elias National Park in 1980. We saw cute sea otters pop up from the water and Arctic terns and puffins. I marveled at the varied shapes and sizes of blue-hued icebergs silently waning in their journey seaward. We were tentative in our approach, as unstable icebergs can "flip" without warning, stimulated by the force of gravity on their changing shape. When an iceberg plunges over, the shift in water can create a huge wave. We didn't want to get swamped!

At the end of the month, we packed up my Ford Bronco, took the ferry to Haines, and embarked on the Alaska-Canada Highway to Anchorage. I wanted so much to get all my things together in one place, unpack, and sit still for a while. Until that day came, I was off on another adventure.

Game Management in Palmer

"We must accept finite disappointment but never lose infinite hope." ~Martin Luther King, Jr.

The Alaska-Canada Highway, initially built in 1942 as a military access route, took us north to Whitehorse, Yukon Territory. We camped at Lake Laberge, a widening of the Yukon River well known to Klondike prospectors and made famous in Jack London's book, *The Call of the Wild.* While there were still late summer flowers by the lake, the snow geese were restless in the cool air. We spent a day touring Whitehorse, then drove north to Dawson City and into Alaska. At Tok, we turned south, heading for Glennallen, then picked up the Glenn Highway into Anchorage.

In September 1978, we rolled into the largest city in Alaska, located at the head of Cook Inlet. The inlet was named after the 1778 expedition of Captain James Cook, who sailed up its waters while searching for the Northwest Passage. Cook Inlet runs through treacherous mud flats composed of glacial silt that are subjected to extreme tides, traveling at speeds of 10-17 miles per hour. Sadly, a few newcomers ventured out on the flats at low tide, became stuck in the mud as the tide bore down, and drowned—a reminder of Alaska's deadly beauty.

In an apartment looking west over Cook Inlet, I gazed at Mt. Beluga, known as "The Sleeping Lady." The spectacular view reminded me of how lucky I was to be in Alaska! I was now a resident, having passed my one-year anniversary, and who knew what lay ahead?

I quickly settled in the new apartment to attend to paperwork regarding my dad's death, like paying his medical bills not covered by Medicare. Directly related to his bills were *my* bills. I needed a job! I drove to the Anchorage ADFG office at 333 Raspberry Road and walked in. The two-story block building housed a large contingent of staff from all divisions. I asked to speak with someone in the Game Division about a job. As I waited in the hallway, a big guy with a commensurately large mustache came out of an office four doors down and ambled toward me. He smiled, extended his hand, introduced himself as Jim, the Regional Management Coordinator, and beckoned me to follow him to his office. Jim seemed kindly and interested as I explained that I had just arrived in town from Juneau and was looking to continue my employment with ADFG. After a series of questions aimed at uncovering my educational background and experience, Jim leaned back in his chair and paused. He told me there might be an opening and to sit right there; he was going to get someone. I waited, hands folded in my lap, holding my breath, trying not to be too hopeful as the minutes ticked by. Then, Jim was back with a man in his middle years with a twinkle in his clear gray eyes.

"This is Jack. He's the area biologist in the Palmer office," Jim said as Jack came in and sat down.

It was my good fortune that Jack came into the Anchorage office the same day as my visit. Jack and I talked for a while. Jack said he could use an assistant in Palmer to help conduct moose surveys during the winter, and he hired me right there as a nine-month seasonal temporary Game Biologist I. I was thrilled! Although the position was temporary, it was a foot in the door, and I would apply for permanent positions when they became available.

I drove from the apartment in Anchorage to my job in Palmer, about 45 miles. Palmer was a small town in the heart of a thriving agricultural community, created during the Great Depression when the federal government resettled over 200 farming families from the hard-hit Midwest to the Matanuska-Susitna (Mat-Su) Valley. While

I felt at home in the Mat-Su Valley, I also enjoyed the advantages of living in a big city. In winter, I skied groomed cross-country ski trails at Anchorage's Kincaid Park. I joined the Anchorage Kayak and Canoe Club and had a ball learning to roll a kayak and upright it again during practice sessions at an indoor pool.

I envisioned that I would eventually relocate to Palmer, buy a little old farmhouse, a fixer-upper I could afford, and happily work on game research and management in Southcentral Alaska. At twenty-six, I didn't like being without roots. I was unaware of political forces rumbling in the background like a dark bank of clouds that would change my fate forever. My future would not lay in Palmer, and as a result, my life became incredibly enriched as well as pummeled. Maybe you can't have one without the other.

The game biologist stationed in Palmer manages game populations (primarily moose) across nearly 17,000 square miles, including the Mat-Su Valley, called Game Management Unit (GMU) 14, with subunit 14A around Palmer. Subunit 14B extends from Willow to Talkeetna. The Palmer game biologist also manages game on the west side of Cook Inlet, called GMU 16, with subunit 16A extending from Susitna north to the border of Denali National Park. Subunit 16B includes river drainages on the west side down to Redoubt Bay in Cook Inlet.

Most of the management area is sparsely populated boreal forest surrounded by mountain ranges. Two major highways, the Parks and the Glenn, and the Alaska Railroad bisect the management area, leading to significant human-moose interactions.

Biologists manage moose for harvest by using aerial surveys to estimate the abundance of animals in a given area and their sex and age composition. Factors affecting their health and survival, such as quality of habitat for food and sources of mortality, are also considered. Harvests are controlled through limits on the number of animals, size, sex, season, methods, and means and tallied through harvest reports returned by hunters.

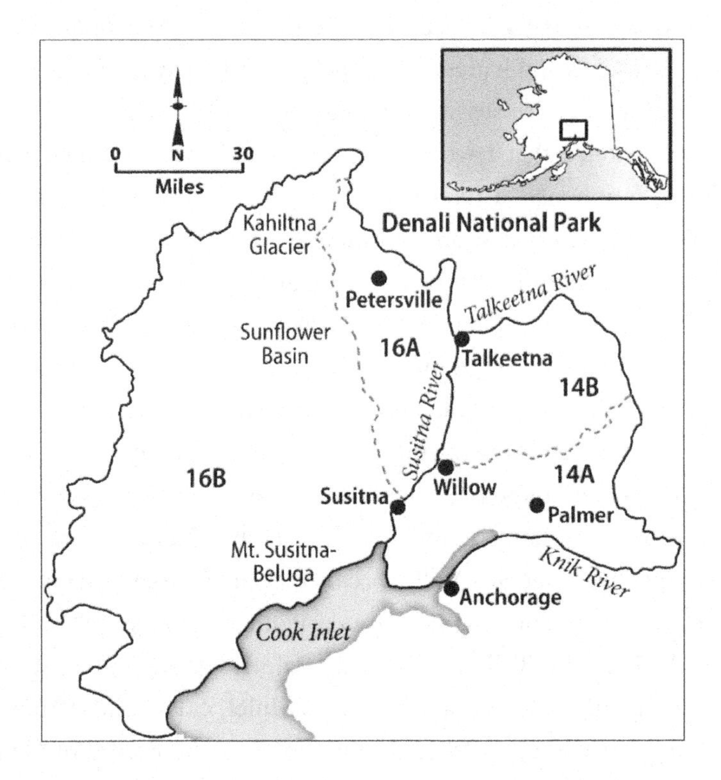

Because of the large area, Jack had an assistant Game Biologist II, Ken, stationed in Talkeetna. As Jack's assistant in Palmer, I had three primary duties: conduct aerial moose surveys, deal with calls concerning human-moose interactions, and write the furbearer report.

Moose are an iconic feature of the boreal landscape, with their distinctive tall legs and long nose. Although solitary by nature, moose groups move seasonally between calving grounds in spring, feeding areas in summer, rutting areas in fall, and overwintering range. Calves born in spring remain with their mothers during winter for protection from predators and to learn where the migration paths are. I conducted aerial surveys of moose on their winter range in early December when they are easy to see against a snowy white background, but before most bulls have shed their antlers.

Survey conditions must be just right to have a good moose count. Ideal conditions are sufficient snow on the ground, temperature above -30°F, wind speed not more than 20-30 miles per hour, and enough light to see. Foul winter weather can delay or cancel surveys, leading to incomplete information for management. In early December 1978, nasty weather delayed my surveys of count areas until mid-December, and I could not fly in some areas at all.

I arrived at the small airport early, with the weak sun just breaking over the horizon. We didn't want to waste the short supply of winter daylight. It was cold. The icy air stabbed my nostrils and my eyelashes frosted from my breath. I dressed warmly, with pack boots, snow pants, a parka, and hand warmers in my pocket, because the inside of the plane was not much warmer than the outside air.

The pilot was already there, fueling the plane. One of the best planes for aerial surveys is the PA-18, a fixed wing, two-seat Super Cub equipped with long-range fuel tanks. It has a slow speed of travel, about 90 mph, which allows enough time to get a good look at the animals, yet the plane has great maneuverability, needed for steep turns at low airspeed. I knew that low-altitude aerial surveys were dangerous, but I was excited about the adventure and the chance to do a job only a handful of chosen people could do.

I usually flew with Rick, an experienced moose survey pilot who knew the country and how to fly to achieve a successful survey. Rick and I looked over the maps of our selected count areas for the day. If he didn't like the local weather in an area, we'd settle on an alternative area or turn back. Rick didn't take many chances.

The weather could change fast and be different between valleys. There were times when snow tumbled suddenly from low clouds, obscuring the wing tips, much less the terrain. Then, we flew blind, and I trusted the pilot to know our exact position in relation to the surrounding hills as he searched for a visible landmark. Other times, turbulence wrested steady control of the plane from the pilot, so I couldn't get a good look at what was beneath us. Aerial surveys attempted under poor conditions resulted in numbers that were minimal. In situations of uncertainty, managers are usually conservative in setting harvest levels.

Count areas were flown in the same manner under roughly the same conditions year after year to generate population trends over time. Transects through the count areas were outlined on the map, and Rick stuck closely to their route so we could determine linear miles flown. As we approached a count area, we looked intently for moose shapes. Rick had to keep an eye on the moose while at the same time keeping an eye on the rising wall of mountain that we might be flying next to, a challenging task.

My survey areas were primarily on the west side of Cook Inlet in GMU 16. When a moose was spotted, Rick dropped to about 500 feet, so I could get a closer look, often tilting the wings to give me a better view. I had a clipboard in my lap with forms and quickly marked what I observed: antlers indicated the moose was a bull, no antlers meant cow; large, medium, or small sized meant mature, yearling, or calf. If a moose was by itself, I could usually assess its sex and age with a brief look, but for a group of moose, Rick needed to make several tight circles over them for me to complete my assessment, leaving my stomach either on the floor or ceiling of the plane, depending on the dive upward or downward. After recording the animals in that spot, I signaled to Rick I was done, and we continued to the next transect until we found another moose.

Aloft, I was entranced by the beauty of the winter landscape. A vast leaden sky edged with a pink polar glow formed a perfect backdrop for trees that had been transformed into confections of white from frozen vapors. Below, little brown moose shapes materialized as we skimmed past a ravine. I felt fortunate to witness such a breathtaking panorama few would ever see.

Fair weather was found in the Peters Hills south of Kahiltna Glacier, where I saw 792 moose or forty-six moose per hour flown. There were thirty-eight bulls per 100 cows, and thirty-three calves per 100 cows. The ratio of bulls to cows indicates hunting pressure in bull-only hunts, while the calf to cow ratio indicates productivity.

Sunflower Basin

In Sunflower Basin, we had good weather and I counted 603 moose, or 128 moose per hour flown, the largest sample ever found for that area. The bull to cow ratio was forty-four bulls per 100 cows, and the calf to cow ratio was nineteen calves per 100 cows.

The survey of Mt. Susitna-Beluga required several days. Initially, the weather was good but deteriorated. I counted 469 moose in twelve hours flown, or thirty-eight moose per hour, less than observed in previous years. Poor survey conditions likely contributed to the low numbers. In 1978, moose overwintering on the west side of Cook Inlet appeared to be in good shape. But the relatively low number of calves observed was a concern for future population stability.

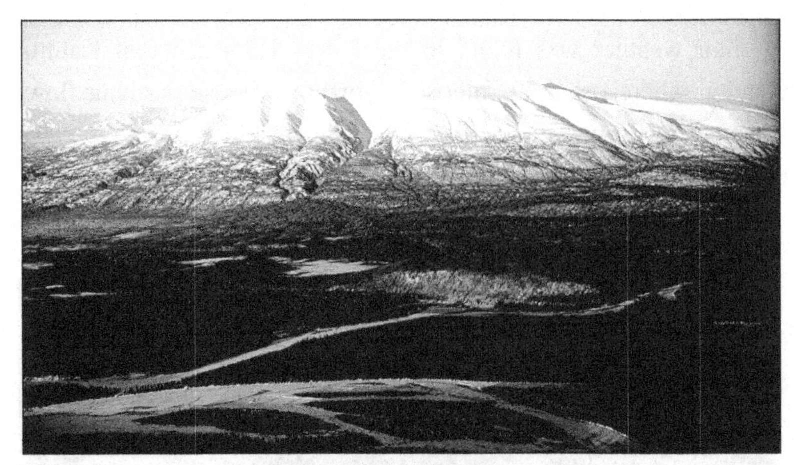

Mt. Susitna-Beluga

I remember one moose we flew over became so agitated by the sound of the plane overhead that he rose on his hind legs and repeatedly stomped the ground with his front hooves, giving me a good idea of how effective moose are at stomping predators or people in the wrong place at the wrong time.

Moose don't like to stand in deep snow because it leaves them vulnerable to wolf attacks, and food is harder to find. So, they seek areas of low snow accumulation or that have been cleared of snow, like roads and railroad tracks. Wintering moose near a concentrated human population leads to significant interactions with often catastrophic results. Adult male moose are large, weighing up to 1,600 pounds and standing 7 feet tall at the shoulder. Car crashes are deadly for both humans and moose. In 1978, there were 108 road-killed moose in subunit 14A. Engineers with the Alaska Railroad try to avoid hitting moose by running pilot cars ahead of trains or sounding a whistle to scare them off the tracks, but despite these efforts, moose deaths from train hits are high. In 1978, 171 moose were killed by trains in GMU 14B. When possible, meat from moose killed by cars or trains is salvaged and donated to charity organizations or low-income individuals.

I answered calls that came in about struck moose. After receiving information on the location and condition of the moose, I pulled on my jacket and drove to the incident with a heavy heart. The moose was almost always dead or dying. It was usually dark or in waning winter light—a dreary scene. If people were injured in the accident, the troopers would have sent them off in an ambulance and dispatched a mortally wounded moose with a shot to the head by the time I arrived. In these cases, I helped move the carcass off to the side of the road, collected information on the moose's condition and sex, and took the lower jaw for age determination before salvagers came.

Collisions with moose continue to plague Mat-Su drivers. As a warming climate creates milder winters, overwinter survival of moose can increase. While an increased abundance of moose is good news, more moose roaming the valley corresponds to more collisions. Biologists attempt to reduce moose-vehicle collisions with targeted hunts in high-traffic areas.

Another non-hunting danger moose face is from dogs. Stray or loose dogs can pack up and threaten moose, similar to wolf packs. Calves are especially vulnerable to being injured and killed, despite the cow's best attempts to defend her youngster. I remember one call from a distraught homeowner who had a cow cornered in her yard, attempting to protect her calf from a pack of stray dogs. By the time I arrived, the calf was bloodied and beyond saving. It made me mad; such needless carnage! I always carried a rifle with me and was within the law to protect wildlife from dogs. I did. The cow stumbled away, bleating for her lost calf, but basically uninjured. With luck, she would survive the winter and give birth again. As for the dogs, they had no collars or forms of identification, so their carcasses were disposed of. A sad outcome for all involved.

When humans are on foot, moose have the upper hand. Moose in winter are hungry and cranky and may charge people for no apparent reason. Moose can knock a person down, then stomp them with their sharp front hooves, similar to their behavior when attacked by a wolf. Nearly every winter, there are reports of moose charging or stomping people waiting by the side of a road or skiers out enjoying a trail. Stomping can result in severe injury and sometimes death, so people are advised to steer clear of moose, particularly when their hackles are raised.

The primary furbearer of interest in GMUs 14 and 16 was the beaver, an animal with a long history of commercial trapping in Alaska. I was tasked with assembling, analyzing, and reporting on beaver trapping between 1969 and 1978. To ensure trapping is sustainable, biologists assess beaver populations by periodic aerial surveys of colonies. Harvests are controlled through bag limits, seasons, area closures, methods, and means. The number of beavers harvested is tallied through pelt sealing requirements. The seal, which is a metal or plastic tag attached to the pelt by a biologist with ADFG, allows the legal transport and sale of the pelt. Residents who keep beaver for personal and cultural use may not seal the pelts, so harvest may be underreported.

From 1969–1978, trappers could take forty beavers annually between February and April in Subunits 14A and 14B; the trapping season was three months longer in GMU 16. As I reviewed the computer printout on beaver harvest, I found coding errors. For example, different streams had the same code. I was surprised to discover that beavers taken along the entire Susitna River drainage, which borders GMUs 14A, 14B, 16A, and 16B, were arbitrarily plopped into 16B, thus exaggerating harvest there.

I wondered if the background story would clarify the coding scheme, so I called the man in charge of furbearer harvest coding, explained the errors I had found, and innocently asked him what his rationale was. He became instantly defensive and chewed my head off for even asking such a question! Clearly, he had no rationale and was converting his confusion into anger directed at me. As a newly-hired person, it was hard to bear being yelled at unfairly by a senior biologist. I told him I didn't appreciate being scolded for trying to understand how to do my job better and ended the conversation on a positive note by murmuring what a challenge it is to develop a complex coding scheme. I decided to retrieve the original harvest reports stored in a closet in Anchorage and compile the data by hand instead of using the computer printout.

Harvested beavers are grouped into four size categories based on pelt measurements in inches, obtained by summing the width and length: kits (53 inches), medium (54–59 inches), large (60–64 inches), and extra-large (65+ inches). A low harvest, mainly of kits, suggests a population reduced in number from overexploitation or other stressors, such as poor habitat. In contrast, the harvest of older and larger beavers indicates the population is abundant and can support continued harvest.

During 1969–1978, the total reported take in GMU 14 was 1,764 beavers and in GMU 16 was 3,056 beavers. Most were in the large and extra-large categories. I concluded that current seasons and bag limits were compatible with beaver populations in both GMUs, and no change in regulations was needed. Due to the frequency of complaints about beavers destroying trees and flooding property from residents in Subunit 14A, I recommended that a beaver management plan be developed and implemented to address human-beaver issues. Since the only colony survey on record was flown in 1971, when ninety-five live colonies were reported on the Susitna Flats, I urged another survey to assess the current abundance and locate dams posing potential problems to homeowners. Finally, I suggested a modification of the harvest coding scheme.[3]

At the office one morning in early January 1979, Jack approached my desk with a glum look. He had bad news about my job. Although he had the money in his budget to keep me on as his assistant in Palmer for nine months, and he really wanted to, a ruling resulting from the re-negotiation of general union employee contracts between the state and the Alaska Public Employees Association limited the length of employment for temporary positions to six months. The association bore down on the state to enforce the six-month employment limit. My job would end in March, not June. I wouldn't be eligible for another temporary position for another six months. I could live on a nine-month salary but not a six-month salary. This strategy of six months on and six months off would put a lot of financial strain on temporary employees and jeopardize the work they were supposed to accomplish.

On the one hand, I agreed with the point of view of the Alaska Public Employees Association. If the state had no limits in hiring a cheap temporary workforce and did not have to pay them leave or retirement benefits, there was no incentive to create permanent positions, thus screwing temporary employees. On the other hand, I loved my job and wanted to keep working, hoping to score a permanent position someday. A public hearing was scheduled in Anchorage about the new ruling, and I attended to give testimony. Worry and tension gave me headaches.

My last day of work in March 1979 was a very sad day. I was bitterly disappointed because my hopes had been dashed. I had hoped to be a game biologist in Southcentral Alaska. I had expected to settle in the farming community of the Mat-Su Valley. Damn the bureaucrats! I left the Palmer office, never to return.

In the weeks following, I felt lost and uncertain about my future. I paced in the apartment and walked the sidewalks. I learned two things: don't expect life to be fair, and disappointment can transform into determination to improve my situation. It was my feelings of disappointment that fanned my willingness to try a new adventure. During a distressing time, I discovered the courage to take a risk and the faith that new possibilities would arise.

I decided to go to Kotzebue for a few months to see new country and look for jobs in the Arctic. Matt had transferred to a new job there so I would know somebody. With no strings holding me back, I reasoned it was good timing for a unique and interesting adventure. I put my car and belongings in storage in Anchorage and booked a flight to Kotzebue. Little did I know what was coming my way.

Kotzebue

"Twenty years from now you will be more disappointed by the things you didn't do than by the ones you did do.... Sail away from the safe harbor. Catch the trade winds in your sails. Explore. Dream. Discover." ~Mark Twain

In May 1979, on the afternoon of my twenty-seventh birthday, I peered out the window as the plane's wheels touched down in Kotzebue, Alaska, a coastal Iñupiat village 33 miles north of the Arctic Circle. While all Iñupiat people are hunter-gatherers, culturally, there are two groups: the people of the sea, who live along the coast, and the people of the land, who live upriver. The first white man to officially arrive in Kotzebue was German explorer Otto Von Kotzebue, in the employ of the Russian government, who in 1816 traded tobacco with the locals. Whaling ships arrived in the mid-1800s. Trade increased, and new items brought cultural changes, especially liquor. The first missionaries, Quakers, arrived in 1897. Soon trading posts, schools, a hospital, and churches were built.[4]

Kotzebue sits on a windblown gravel spit at the end of Baldwin Peninsula that supports only sparse tundra. As I stepped out of the airplane, my impression of the ecology was overshadowed by the sounds of ice floes loudly splitting and cracking in Kotzebue Sound. It was spring, and breakup of the frozen sound commanded my attention.

I moved to Northwest Alaska out of curiosity but mostly because I hoped to get another job with ADFG. My presence in Kotzebue held a bit of déjà vu because twenty-three years earlier, my father had been in Kotzebue on a photography expedition.

Knowing that my steps were following his gave me some relief from his loss in my life.

My arrival in Kotzebue coincided with the swirling of political forces directing changes in land conservation and regulation of subsistence hunting and fishing, adding a sharp edge of anger and uncertainty to the atmosphere. In 1971, the Alaska Native Claims Settlement Act resolved long-standing native land claims in Alaska and stimulated local economies by creating regional corporations. A small section of this bill, known as d-2, allowed the federal government to withdraw from development millions of acres in "conservation areas." Congress dawdled over section d-2 for seven years until President Jimmy Carter, in 1978, used the Antiquities Act through Executive Order to designate 56 million acres as National Monuments. Areas designated near Kotzebue included Cape Krusenstern, the Noatak and Kobuk River Valleys, and the upper Noatak and Kobuk Rivers, which in time, came to be part of the Gates of the Arctic National Park and Preserve.

Alaskans were caught off guard and were furious with the perceived federal land grab, prompting Congress to pass the Alaska National Interest Lands Conservation Act in 1980. A section of this bill, known as Title 8, allowed the federal takeover of subsistence hunting and fishing management on federal public lands. Alaskan biologists were initially confused about procedures and roles in the dual management of animals that wandered across state and federal lands. Additionally, native and rural subsistence rights trumped all other harvest opportunities, unintentionally spawning discord and rivalry between different user groups vying for limited resources.

When I arrived in 1979, the population of Kotzebue was about 2,000. The main thoroughfare, Front Street, is built right on the beach and follows the lay of the sound. Old houses line the gravel street; most are no more than shacks, their gray wooden siding weathered and bent. During breakup, the sound ice cracks and moves, opening and closing leads.

Open water leads in Kotzebue Sound in spring

At that time, a custom of the locals was to walk the short distance to the sound and toss a trash bag or honey bucket contents onto the ice, which was, after all, going out to sea. The reasoning was that the ice would carry the trash into the ocean and thus save the trouble of taking trash to the local dump. As I walked down Front Street on my first morning, I saw colorful arrays of garbage strewn on the beached ice floes, waiting to sink. Later in summer, storms washed up reminders of odds and ends left by the residents of Front Street.

The ADFG office was in a building on Front Street, which also housed the U.S. Post Office. To get to ADFG, you had to walk through the post office lobby and down a hallway. The lobby was a convenient shelter for drunks to sleep off their night's excesses, and it was common for me to negotiate my path through and over inebriates to get to the office door. One of these men I became acquainted with was named Frank. He smiled a lot. One of Frank's favorite pastimes in summer was to get his photo taken by tourists and tell them tall tales of Arctic life.

Alcoholism in the village was a problem and led to domestic violence issues, as I came to see first-hand. I rented an apartment in a new complex with a locked security door. The apartments had thick carpet and nice furniture for only $550 a month. One night, at 10 o'clock, a loud pounding suddenly started on my door, then shouting. The hollow wooden door began to splinter! I rushed to the phone, and as I dialed 911, a drunken woman bashed through what remained of the door and came at me with hands outstretched, like she was going to choke me! I was totally unprepared for this encounter. Fortunately, the woman realized she was in the wrong apartment and turned and left as suddenly as she had appeared. The police later told me that she was after her boyfriend's new ladylove for revenge. Beatings, fights, and abuse fueled by alcoholism were undercurrents to life in Kotzebue. One escape was to go upriver.

Three rivers drain into Kotzebue Sound: the Noatak, the Kobuk, and the Selawik. For centuries, these rivers have been important routes for trade between coastal and inland villages and remain primary transportation routes for area residents. In summer, boats are an essential mode of transport; in winter, snowmachines and dog sleds are used on frozen rivers.

In my first week in Kotzebue, I quickly learned a skiff is a vital asset for leaving the windswept peninsula and its troubles behind. Upriver travel promised scenic tree-lined hills rich in fish and game and a way to refresh my spirit. I went to K.C. Company, a major chain of stores in the rural Northwest, and purchased a 16-foot aluminum Starcraft skiff with a 35-hp Evinrude outboard engine for $2,000. Like other Kotzebue residents, I moored the skiff on the beach tied to a large piece of driftwood.

Tides were small, often less than a foot between high and low tide, so I could easily drag the skiff down to the water when I was ready to cast off. In early June, even though ice floes still float in the sound, boating season begins. Sunrise is at 2:30 a.m., approaching the 24-hour daylight of the Arctic. The temperature is above freezing but can still be frigid with the wind blowing across the ice.

I was eager to explore the country with new-found friends. In Kotzebue, there were two types of residents: locals and transients. Transients mainly consisted of white folks who rotated in and out of the village every two or three years and were essential to the modern functioning of life. Groups of transients included U.S. Public Health Service members who staffed the local hospital, teachers hired by the Northwest Arctic School District, and miscellaneous federal and state employees such as air traffic controllers and biologists. People gave varied reasons for committing to a tour of duty in the Arctic, but most were based on ready employment as well as a chance to have a unique adventure. Because an Iñupiat village on the edge of the Arctic coast was so different from anything we had experienced before, most transients tended to form a close-knit bond of easy friendship, perhaps as a source of familiarity in an unknown land. Within a few days of my arrival in Kotzebue, I became a member of several groups of friends. A few months later, it was my turn to welcome new transients into the ever-revolving circle of volleyball, softball, and basketball teams, dinner parties, and camping.

The first weekend in June 1979, I took my skiff on its maiden voyage with new camping buddies. We pushed through the ice slush in Kotzebue Sound and navigated around the tricky Noatak River delta with a few false starts until we found the main river channel. By departing Kotzebue at midnight, a soft twilight time, we were on the Noatak River by sunrise. I was thrilled to be at the controls of my Starcraft, learning to interpret signs in the swirls and currents of the water and testing my skiff's abilities and limits. Once past the ice floes, the land mass upriver warmed the ambient temperature. Myriads of waterfowl rose around us on the journey. Gaunt, gray hills hugged the river valley, with tree lines halfway up, leaving the tops of the hills looking bald in their tundra caps. Sweeping clouds cast flickering shadows, inspiring awe and respect for this harsh and primitive beauty. Setting up camp in a stiff cold wind was a challenge but being in the Arctic country in early spring was invigorating.

Two weeks after my arrival in Kotzebue, I had an opportunity to assist reindeer herders employed by the Northwest Arctic Native Association in rounding up reindeer at Kiwalik, on the Seward Peninsula, for antler harvest. Reindeer were imported to the Seward Peninsula from Siberia in the late 1800s. Following a profitable period through the 1930s, the husbandry of reindeer declined. The native association began the reindeer operation with 800 animals in 1976, and three years later, the herd numbered over 3,000 animals. Initially, the reindeer were intended to supply meat to area villagers to offset periods when caribou numbers were low. However, a more lucrative venture arose with the harvest of wet velvet antlers sold to Asian buyers. The antlers were thought to have medicinal or aphrodisiac powers in the Far East and commanded a high price. Since reindeer shed their antlers and grow new ones annually, the animals were not killed during the antler harvest.

Roundup in 1979 was accomplished with a helicopter. The reindeer were herded through a pass and across a shallow stream into a large holding pen. My job was to capture and tag fawns. In the milling herd, fawns became separated from their mothers, and both mother and baby bleated incessantly, calling to find each other. Miraculously, babies and mothers reunited at the end of the day and were released onto the tundra to roam freely until the next roundup time.

Reindeer roundup at Kiwalik

Male reindeer were separated from the does and fawns and prodded into a squeeze chute, where a crew of men went to work harvesting antlers. In June, antlers are "in velvet" and enriched with a blood supply that nourishes the growing antlers. Asian buyers prize antlers engorged with blood, so the antler harvest was initiated in June to obtain a high market price for their product. At that time, the antlers were sold for $25 per pound.

Once the male reindeer was subdued in the squeeze chute, antler harvest occurred in the following manner. Man #1 quickly wrapped a rubber tourniquet around the base of each antler to stem squirting blood. Man #2 cut off each antler above the tourniquet with large pruning shears. Man #3 poured a white powder onto the stump, which helped staunch bleeding and also acted as an antibiotic. Man #4 tossed the harvested antlers onto a big black tarp. Man #5 opened the chute door to release the reindeer back into the pen. This procedure was repeated throughout the day. At the end of the day, reindeer whose antlers had been harvested were separated and herded into the squeeze chute for a second time. The men removed tourniquets, the animal was quickly given a health check, and then shooed out of the pen onto the tundra.

Korean men who had purchased this lot of harvested antlers were on site that day to oversee the operations. The sight of squirting blood from the antler stump inspired the Koreans to lean over the reindeer in the squeeze chute and joyously gulp the blood with open mouths. I was absolutely stunned at the spectacle. Not only did I feel squeamish to watch them drink fresh blood, but more importantly, a significant portion of reindeer in Alaska harbor brucellosis, a disease readily transmitted to humans through uncooked meat and contact with blood.

I was impressed with the reindeer enterprise as a means to stimulate income and employment for native people in an economically depressed area of Alaska. The animals were handled and cared for in a professional manner. I was one of only two white people on-site during the operation that day, and I think most of the men appreciated our help. I heard only one man ask aloud what white people were doing at Kiwalik. "They aren't wanted around here!" he said. For the first time in my life, I became aware of what it felt like to be in the minority, and my presence resented because of how I looked.

While the first weeks of living in Kotzebue had brought incredible adventures, the lack of a permanent job as a biologist weighed heavy on my mind. I was aware of expenses eating away at a dwindling reserve. Paying my living expenses rested solely on my shoulders. I went to the ADFG office and introduced myself to the lone biologist, Dave, who was the Game Division's area management biologist for GMU 23. Dave was a tall, laid-back Swede who had lived in Kotzebue for several years, slipping comfortably into the community. I learned biologists had a particularly challenging time building rapport and trust among the Iñupiat in the Northwest Arctic, who harbored suspicion and resentment stemming from Territorial days.

I was told the following background story. In the 1940s, the Western Arctic caribou herd, a primary source of nourishment to the Iñupiat for thousands of years, had significantly declined in number. At that time, intensive hunting by natives was cited as a primary cause of the decline.[5] Out of concern for dwindling caribou numbers, federal biologists imposed limits on hunters, who resented regulations on their traditional ways. The feds traveled to villages to explain how important it was to the herd's health to decrease hunting mortality, warning that those who violated federal law could be taken to court and their rifles confiscated. Forty years later, the threat of having their rifles taken from them was still fresh in the minds of locals, who did not distinguish between state or federal agencies—government biologists were all the same. Thankfully, in 1979, the Western Arctic caribou herd was in fine shape.

Dave told me that the FRED and Commercial Fisheries divisions were sending biologists for fisheries research, and he would pass along my name to them. A few days later, I got a call. A biologist named Jim with FRED was arriving in Kotzebue later that week and wanted to contract my boat and services. I took Jim on two charters. He paid me $10 per hour for the boat charter and $14 per hour for my services as a biologist. On the first trip, we motored along the shoreline of the upper Baldwin Peninsula and seined for juvenile chum salmon. On the second trip, I took him 36 miles up the Noatak River, past the lower canyon, mapping tributaries and seining the shoreline for juvenile chum salmon.

Two weeks later, Jim called to offer me a job as a part-time, nonpermanent Fishery Biologist II with FRED. I was disappointed the position was not permanent, but it was a beginning. I had noticed Jim was precise in his scientific work and admired him for that, so I thought he would be a good boss. Jim said he had to fight the Alaska Public Employees Association to hire me as a nonpermanent employee because I had only been out of work for three months, and the association mandated a six-month break from work for positions designated by the state as "nonpermanent."

As a bonus, Dave said I could work for him in the winter as a seasonal, part-time, nonpermanent Technician I. Having a Master of Science degree, I knew the Game Division would underpay me for the services I was about to perform for them. Still, I was grateful for work in the frozen North, where few employment opportunities presented themselves. I was also grateful to be a biologist in new and mysterious country! I reminded myself that I shouldn't want the stars and the moon, or I'll never be happy with what I have at any one time. Between the two jobs, my financial worries abated, and I looked forward to the exciting work ahead.

Chum Salmon Research

My Fishery Biologist II job with FRED lasted from June 1979 through December 1980 and consisted of two parts: research the early life history of chum salmon, also called "dog" salmon; and investigate locations throughout the Kotzebue Sound area for a suitable place to build a chum salmon hatchery.

Commercial chum salmon fishing in Kotzebue began in 1909 when a trading post purchased over 20,000 pounds of salmon from residents for resale to prospectors at 5 cents a pound. After a lengthy hiatus, buyers returned, and a commercial fishery was organized under state management in 1962.[6] The year I arrived in Kotzebue, the total commercial catch was 141,545 chum salmon, with 181 boats fishing. Fishermen received 80 cents a pound. The estimated ex-vessel value was about $990,000, contributing significantly to the local economy.

There are two major runs of chum salmon in Kotzebue Sound: the smaller Kobuk River run peaks in mid-July, while the larger Noatak River run peaks during the first half of August. The opening date was July 10 to reduce harvest of the smaller Kobuk run. Managers monitored harvest, catch-per-unit-of-effort, aerial survey data, and test fishing results to decide weekly fishing times.

Because the commercial catch had fluctuated widely since the onset of state management, ADFG wanted to investigate whether a hatchery would moderate chum salmon production, thereby providing fishermen with a more stable income. Hatcheries virtually eliminate freshwater mortality, a contributing factor to adult salmon production. However, research on the early life history and migration of chum salmon in the Noatak River and Kotzebue Sound was needed before any serious talk of building a hatchery. That's where I came in.

Previous work on chum salmon in Kotzebue Sound had been confined to adults. Little was known about how chum salmon endured the arctic climate during their larval and juvenile stages. My job was to investigate the timing and locations of chum salmon spawning, development rates of eggs and alevin, food availability for emerging fry, and environmental factors affecting survival. I was to monitor the timing and duration of the outmigration of fry and follow their growth and feeding in nearshore habitats in the ocean. I would determine the availability of zooplankton as a food source from plankton net samples. Migration routes were to be mapped and oceanographic data collected. It was a groundbreaking project!

During the summer of 1979, I used my little skiff to explore nearshore waters for juvenile chum salmon as they spilled out of the Noatak River channel and into Kotzebue Sound. At this point in my boating career, I decided to take a small engine repair class. I kept spare parts and tools in ammo boxes in the skiff. There were several instances when I had to replace the propeller, change the spark plugs, unclog the fuel filter or unblock the water intake. I even had an extra line for when the flywheel cord broke. I learned that being prepared and improvising were the keys to a successful outing!

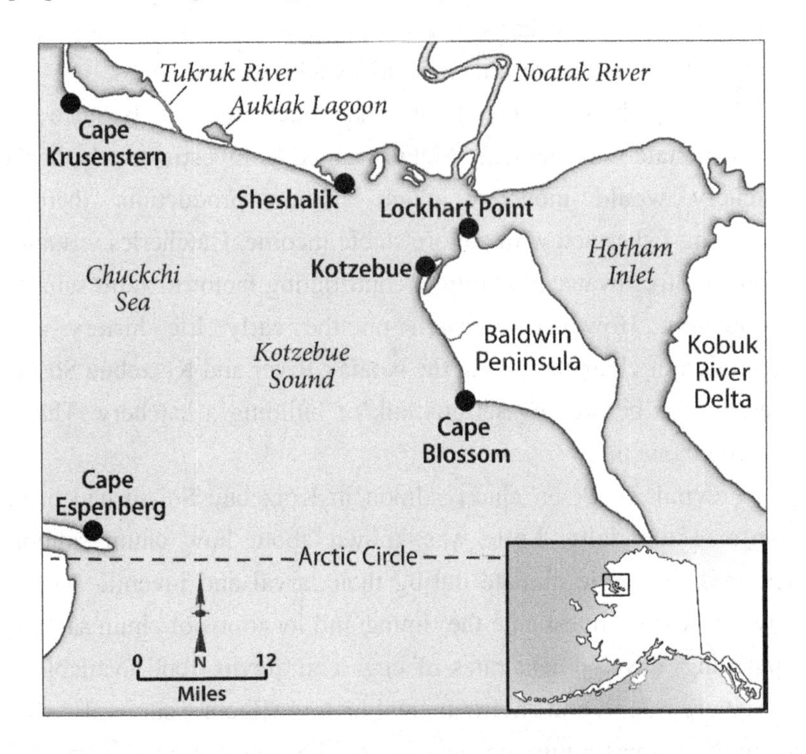

On a bright summer day, I set out on my first solo mission as a Fishery Biologist II in Kotzebue, casting my skiff off and setting a course along the northern tip of the Baldwin Peninsula toward Lockhart Point. In the skiff were collection bottles, maps, notebooks, a stake, and small mesh beach seines: a 20-foot net for pulling by hand, and a 150-foot net for deeper hauls with the boat.

Preparing to seine for fish with my skiff at Lockhart Point

While my college education did not give me many practical field skills, it gave me confidence in problem-solving and resourcefulness. I figured out how to make a beach seine haul from a boat with just one person. Arrange the beach seine on the bow with floats on one side and weights on the other, drive a stake into the ground close to the waterline, tie the end of the weighted seine onto the stake, then slowly back the boat in a semi-circle as the folds of the seine fall off into the water. The tricky part was drawing close to shore prop-first. In fast order I had to shut off the engine, flip the prop up before it hit gravel, jump into the water holding a line to the boat and a line to the seine, and pull boat and seine to shore, careful that the weighted line stayed on the bottom so the fish couldn't escape. For shallow hauls, I just walked the seine into the water from its staked end in a semi-circle and back into shore.

Captured fish were quickly identified, counted, measured, and released. A sample of chum salmon was preserved in a formalin solution for later stomach content analysis in a lab. I meticulously recorded the location of each haul on a map, the time of day, weather conditions, water temperature, and catch-per-unit-of-effort.

As summer progressed, my boss found money so I could periodically hire help with fieldwork. A sonar project on the Noatak River provided a technician on occasion. The addition of Janelle to the seining operation was most welcome.

Janelle and I use a beach seine to capture juvenile fish

We examine our catch

A second person allowed me to sample on days when the water was choppy. Potential predators of chum salmon fry, such as sheefish, were purchased from fishermen to examine their stomach contents. Predators caught during seining were also examined for stomach contents.

On calm days I motored 9 miles across the sound to Sheshalik (Sisualik), a grass and wildflower-covered spit jutting into the Noatak River delta, which is home to subsistence families in spring and summer. I seined along the outer coast of Sheshalik as well as the inside waters of the lagoon. At Sheshalik, I met Bob and Carrie Uhl. Bob transplanted himself to the Arctic from California in the 1940s and married an Iñupiat woman named Carrie. By melding the best of the two cultures, they supported their family in traditional ways.

I remember conversations with Bob and Carrie as we sat on comfortable cushions inside their canvas wall tent, with their little black dog, Snoopy, resting at Bob's feet. The tent walls were lined with shelves made from wooden planks and held assorted dishes, books, and candles. There was always kindling at the base of the wood stove to keep the pot of coffee hot. In one conversation, Bob shared his thoughts about the emerging search by Northwest Alaskan native communities for a revitalization of spirit and reverence of nature. He mused, "It strikes me as particularly meaningful at this time. The missionary drew Eskimos away from nature worship and shook their respect for natural resources. Now, there is more concern for the land and a reassessment of values."

Bob was a self-taught naturalist, and I enjoyed our conversations as he explained the cycle of fish and wildlife on Sheshalik that provided sustenance: bearded, ringed, and spotted seals, beluga whales, and an occasional walrus arrive as the spring ice moves out; chum salmon return in the ice-free months; and berry-picking and migrations of caribou begin in fall. One October, thousands of large, adult hermit crabs washed ashore at Sheshalik.

Bob theorized the die-off was caused by the sudden flushing of freshwater into the marine waters, resulting in a "freshwater shock" to the crabs. He reported, "We ate all the crabs we could hold for a week!"

In turn, I surprised him with the data I was collecting. For example, large peaks in zooplankton density were observed in Kotzebue Sound in July (7,400 organisms per cubic meter) and again in late September (4,000 organisms per cubic meter). Despite the abundance of zooplankton as a food source for juvenile chum salmon rearing in the nearshore waters of Kotzebue Sound, their stomachs were full of insects!

Carrie introduced me to the delicate taste of dried oogruk (bearded seal) dipped in whale oil, which I preferred to muktuk (whale) and kok (dried fish). Seal meat, cut into strips and hung on wooden racks, dried quickly in the constant coastal wind and summer sun. Dried seal meat tasted like dried beef, made rich and fatty from whale oil. In the years after I left Kotzebue, we corresponded, and I treasure the letters I received from Bob and Carrie, highlighting the seasons, animals, and events in their lives at Sheshalik.

In the fall, I shifted my investigations up the Noatak River to chum salmon spawning grounds near the village of Noatak. Noatak people moved between their village, perched on the western shore of the Noatak River 50 miles upriver from Kotzebue, and the coast. Villagers hunted marine mammals along the coast in spring; netted salmon and Dolly Varden from the river in summer; and hunted sheep, bears, and caribou inland in the fall.

In 1979, about 260 people lived in Noatak. There were no hotels. When visitors arrived, they were directed to the school principal, who allowed them to sleep on the gym floor when classes were not in session, and that's where I slept on my first visit. I became acquainted with the teachers, transients who enjoyed the Arctic lifestyle so much they had remained. I was invited to stay in their homes, elevating my sleeping status from a floor to a couch.

My tasks in Noatak were to monitor the water temperatures in the river, locate chum salmon spawning sites in warm spring areas, establish sites to monitor egg and alevin development through the winter, and study the feeding habits and migration routes of emerging fry in the spring.

I became known around Noatak as "The Fish Lady" as I always talked with people about salmon. As I walked through the village, people waved to me from their doorstep with broad smiles, calling, "Hey, Fish Lady, come in and have a cup of coffee!" A traditional gesture of village hospitality was giving coffee, and the polite acceptance of coffee was expected. By the time I walked through the village and accepted six invitations from friendly village folk to have a cup of coffee, I was buzzed and needed to pee. As a woman, I felt I had an advantage in the villages because I was perceived as less threatening and authoritative than a male biologist. People were inclined to share information with me. As a result, I developed cooperative relationships with the locals that helped me meet my research objectives.

Sometimes, I motored upriver in my skiff, but usually, I boarded a commuter flight from Kotzebue to Noatak. The planes were either de Havilland or Cessna aircraft outfitted with wheels in summer and skis in winter. They carried passengers, supplies, and mail. We didn't fly high, and I enjoyed looking down on the tundra.

I recall one plane trip in particular. The pilots up north were usually young and just beginning their careers. One young pilot was on his first trip ferrying people from Kotzebue to Kivalina, then on to Noatak. There were seven of us on this winter trip. After safely touching down in Kivalina, shuffling off supplies and a few passengers, we taxied down the frozen runway adjacent to the village cemetery. The cluster of wooden crosses jutted starkly from the glittering field of windswept snow.

Wooden crosses mark graves next to the Kivalina runway

The white landscape appears featureless in winter until you become acquainted with it. After being in the air for ten minutes, the pilot called out over his shoulder, “Does anyone know where Noatak is?” One of the old Iñupiat men crept up to the cockpit and stayed with the pilot, crouched down on his heels. The old man directed the pilot to fly over the hills until he could drop into the Noatak River drainage, then follow the river to the village. The passengers didn’t think much of it; they were used to transient pilots new to the area.

I hired a Noatak Iñupiat man named Chester to transport me to sampling sites on the Noatak River. Chester was married to Carrie Uhl’s sister, Martha. As Chester and Martha came to know me, they invited me to stay with them and their kids in their small wooden-framed house, overspilling with sleeping bags and clothes in the cozy living room. I was grateful to be included in the family and to share their meals. In the morning, Martha prepared breakfast by filleting a Dolly Varden with an ulu, a curved “women’s knife.”

After she cooked it on the stove, she placed the fish on a large platter, and we all used our fingers to break off savory flakes of meat. Sometimes in the evening, we had caribou stew for supper.

By boat, Chester and I located chum salmon spawning areas during peak spawning time—mid to late September. I flagged twelve sites for return investigations throughout the next nine months. On my visits to the study sites, I recorded water levels, surface and intragravel temperatures, dissolved oxygen, conductivity, and pH. I noticed the spawning areas were distributed along a line of upwelling warm springs in eastern sloughs of the Noatak River. Female chum salmon selected places in the river of a specific gravel size and looseness to lay their eggs. Too much sand and mud blocked the circulating water and oxygen in intragravel spaces, which is necessary for egg survival. I watched as a female dug a hole in the loose gravel with her tail as a male hovered above. Dolly Varden and Arctic grayling scooted around the mating pair, hoping to snatch eggs carried away by the current. After fertilization, the female swept gravel over the eggs to protect them from predators.

As fall waned, water levels in some spawning sites dropped by 8 feet, leaving redds high and dry. The eggs in these redds died. At times of low water, available spawning sites become limited, so females will superimpose redds on each other, displacing and exposing previously laid eggs.

When ice covered the river, Chester transported me to the study sites via snowmobile. I sat in an old wooden sled towed behind his sno-go. One day as Chester negotiated a trail along ice bridges that were suspended across open water leads in the Noatak River, I shouted, "Chester, how do you know that the ice bridges will hold?" He turned his head and replied with a toothless grin, "Sometimes, I don't!"

Steam rises at a study site on the Noatak River

As snow accumulated over my sampling sites, gray gravel was remodeled into a crystal kingdom with a mysterious aura caused by steam rising from warm water upwellings. I discovered the chum salmon spawning grounds were characterized by water temperatures 5°F higher than temperatures in the main river. Intragravel temperatures dropped from around 48°F in September to 35°F in April. Since water temperature determines the rate of salmon embryo development to reach the hatching stage, calculated in temperature units, higher temperatures predicted faster development rates. Most chum salmon eggs and alevins living in these warm spring gravels acquired enough temperature units (about 1,130) to grow and emerge in sync with the spring appearance of insects, such as stonefly and cranefly larvae, necessary prey for juvenile chum salmon growth and survival in-river.

However, some females selected gravel sites with temperatures too low, so their eggs were only exposed to 650 temperature units and thus were delayed in their development. Other females chose gravel sites with temperatures too high, so their eggs developed too fast. I saw one pool of water with alevins emerging from the gravel in January! There was little for them to eat, as insect larvae are not observed until March, so these young chum salmon died of starvation. The gut fullness of chum salmon fry peaked during their outmigration in May and June. The availability of suitable spawning areas, and the choice by adult females as to where to deposit their eggs, appeared to be the factors limiting chum salmon production in the Noatak River.

During my second summer of juvenile chum salmon investigations, FRED purchased a 17-foot Boston Whaler with a 75-horsepower engine so I could safely explore farther along the coast and in offshore waters. The boat was ordered in the winter of 1979 and shipped on the first barge in 1980. The Boston Whaler has a deep V-bottom to cut through waves and is very stable. It is made with a polyurethane foam core coated with fiberglass and is virtually unsinkable. I could really go places now! At twenty-eight, I was the master of my decisions and time. I loved pushing the throttle open and skimming the small chop with the smell of the sea and a light spray brushing the sides of the boat. On these days, I was glad to be an Alaskan biologist, living a life better than I could have ever imagined.

Beach seine hauls were conducted along the western shore of Baldwin Peninsula as far south as Cape Blossom and along the shore of Sheshalik Spit as far west as Cape Krusenstern. I added a longer beach seine (200 feet in length) as well as tow (26 feet long with a 9-foot square opening) and trawl (20 feet long with a 16-foot square opening) nets to my sampling gear, all with small mesh to catch juvenile chum. The tow and trawl nets allowed the sampling of offshore waters, especially in the Noatak River channel. These nets were pushed off the stern and towed behind the boat for a given period. I used a flow meter to record the distance of each tow. Surface plankton nets were also towed behind the boat, and samples preserved in buffered 5 percent formalin for later analysis of zooplankton density.

For most of the second summer, I was by myself when conducting research; however, it was not originally intended to be that way. My boss had budgeted for me to be able to hire an assistant, and I advertised for a local hire. An Iñupiat teenager, shy but keen on fisheries work, applied for the job, and I was delighted to have him on board. I looked forward to teaching him about the scientific study of fish. Sadly, twenty-four hours after I hired him, he told me he couldn't work for ADFG because his friends were angry that he had taken the job and wouldn't speak to him. In 1980, there was general antagonism from natives against ADFG. Some native peoples resented ADFG's management style, methods, and outcomes. Although I felt accepted and even welcomed in the outlying villages, in Kotzebue, the antagonism against ADFG, and to some extent white people, had a sharper edge. This resentment was expressed in the ongoing lobbying of Congress to pass the Alaska National Interest Lands Conservation Act that would expedite subsistence hunting and fishing rights on federal public lands, thus sidestepping objectionable state regulations. It was an unfortunate period in the history of ADFG. In later years, ADFG became more inclusive of stakeholder preferences, and in my tenure, native peoples became ADFG researchers and managers. But, returning to my dilemma in 1980—lack of field assistance.

I cobbled together a solution of three working parts. I continued my partnership with Commercial Fisheries Division staff at the sonar station on the Noatak River; I could periodically hire them when they were available. Additional Commercial Fisheries researchers arrived in Kotzebue in the summer of 1980 to conduct test fishing for herring in Kotzebue Sound. Reports of native catches of herring had prompted ADFG to investigate whether herring were abundant enough to start a commercial herring fishery. I told the new arrivals, Craig and Joe, two boats were better than one because if one broke down, there would be a way to get back to town. We quickly joined forces. When they set off in their skiff, I accompanied them in my Boston Whaler. They helped me with beach seine hauls for juvenile chum salmon, and I helped them with their gillnets for herring. And finally, my boss enjoyed getting out of his office, so he occasionally flew into town to spend a week helping me conduct research.

One research trip in early 1980 brought unanticipated complications and heartache. While the herring crew and I were investigating fisheries resources south of Cape Krusenstern, we came across a dead whale washed up on the beach. To understand the implications of the dead whale requires some background information. In spring, bowhead whales migrate along the Alaskan Arctic coast from their overwintering grounds in the Bering Sea to summer feeding grounds in the eastern Beaufort Sea. For thousands of years, hunters from coastal villages, including Kivalina, had hunted the bowhead, a subsistence activity protected under the 1972 Marine Mammal Protection Act and regulated by the International Whaling Commission.

Because neither the abundance of Bering Sea bowheads nor their harvest was known, biologists with the National Marine Fisheries Service began field research in 1976. When hunting whales, a strike from a harpoon does not necessarily result in a landing. Biologists became concerned about the number of bowhead whales struck and lost by hunters—a total of 111 strikes in 1977. In response, the International Whaling Commission instituted a small harvest quota in 1978–1979 of eighteen struck or twelve whales landed, whichever came first, to meet the subsistence and cultural needs of Northwest Arctic Eskimos.[7,8] Alaskan Eskimos disagreed with the harvest limits and were not eager to volunteer information on strikes. Hunters from Kivalina had not successfully landed a bowhead since 1977. If the whale we found was struck and lost, it would count against their harvest quota.

I finished the day's research, moored my boat, put away sampling gear, and reported the whale sighting to my superiors, Dave and Jim. As far as I was concerned, I had done the proper thing. The next morning at the ADFG office, the newly installed Subsistence Division hire, who had just moved to Kotzebue, angrily approached me. Glen said I was wrong to have reported the whale sighting. Glen was a 200-pound white man, eager to convince locals his new role in the Subsistence Division was to act as their protector and advocate.

I was shocked Glen wanted me to lie! In college, it was drummed into me that science is based on truth, and I was obligated to uphold the scientific profession's integrity. Attacks on scientific facts set the stage for more significant problems—the outright rejection of science in favor of personal beliefs. In this case, Glen wanted me to ignore or cover up what I had found to create what he viewed as a preferable outcome for locals. Despite his menacing stance, I stood my ground.

Just then, a biologist with the National Marine Fisheries Service, Steve, walked into the ADFG office, looking for me. He wanted to know the location of my whale sighting and asked if I could take him to it. Glen became so agitated he lunged at me to silence me, picked me up, and slammed me against the wall! Startled, Steve stepped between us and pushed Glen away from me. Steve told him to calm down and escorted him down the hall. I leaned against the wall, reeling, not so much from the physical damage (although my arm did have a bruise), but from the unexpected and unwarranted attack.

I found my breath and moments later walked with Steve out of the ADFG building to my boat. We cast off, and I retraced my route to the beached whale so Steve could take photos and tissue samples. Eventually, the whale was identified as a juvenile gray whale that had probably strayed off course. The poor creature was determined to have died of natural causes and just happened to wash up on the beach where I was sampling. Gray whales are primarily sighted in the southern Chukchi and northern Bering seas and are not normally found near Kotzebue Sound.

Although the whale was not counted against the harvest quota, Glen insinuated to locals I was not "on their side" and advised them not to cooperate with me. Sometimes, when people are insecure, they put others down to make themselves feel superior, and that's what Glen was doing. Being undermined by an ADFG employee made me despondent. My time developing a rapport with locals was being put at risk! However, I would not have changed my actions because I did what I believed was right. My boss filed a complaint with the state about my mistreatment by Glen, who eventually left the Subsistence Division.

Another unsettling event happened that summer of 1980 while sampling the waters off Cape Krustenstern. The closest you'll come to living in prehistoric times is walking the coast of Cape Krusenstern, a lonely and exposed expanse of seacoast, with the infrequent Iñupiat hunter leaving no trace after traversing the winter icepack looking for bearded seal in open water leads. Occasionally, wooly mammoth fossils wash onto the ancient beach, reminding the visitor of who walked there before.

It was in this area of the Northwest Arctic a member of the herring crew decided to set nets for test fishing. During a fateful week in July, I decided to go along. We would use his skiff to test fish for herring as well as conduct beach seine hauls for juvenile chum salmon.

Under a bright and clear sky, we packed the 24-foot-long skiff with supplies for a four-day stay and pulled anchor. We eased northwest across the Noatak Channel that emptied into Kotzebue Sound, turned westerly, and entered the waters of the Chukchi Sea as we kept the coastline in sight on our right. The seas quietly rolled with no whitecaps. We mapped our location from landmarks: first passing Sheshalik Spit, then the freshwater stream emptying into the sea from Aukulak Lagoon, and finally, Tukruk River, which drains Krusenstern Lagoon. We had arrived at the southernmost line of our established sampling area. Joe eased the skiff into the mouth of the Tukruk River, and we motored upstream a short way to set up camp.

The noise from the engine flushed golden plovers and whistling swans as we nosed up the river. After we set up camp, we used the remaining light for beach seine sets. The hauls were rich in species: smelt, cod, flatfish, nine-spined sticklebacks, shrimp, and sturgeon poacher, but no chum salmon. We decided to set the herring nets the following day.

In the night, the wind picked up. The next morning the sky was leaden, the air heavy, and the ocean swells bigger than the previous day. There were no whitecaps, so we thought the conditions were okay for setting the nets. Joe was at the helm as the skiff moved out of the mouth of the river into the ocean. A few minutes later, we were in the swells. It's funny how deceptive the ocean can look from land. The swells were larger than we had expected, maybe 8 feet high and close together. Trying to set nets off the bow would be difficult, so Joe quickly changed his mind and made a terrible mistake. He turned the skiff 90 degrees to return to the beach. Broadside to the swells, the skiff was instantly swamped and flipped upside down.

It was that fast, in a fraction of a second, and I was underwater with the skiff on top of me, frantically kicking away from the gillnets as they swirled around my legs. The underwater froth churned nets and gear violently around me and clouded my vision. The searing cold water choked the breath out of me, and its weight pulled me down. Two imperatives drove my terrified struggles—to get away from the skiff and head to shore. Thankfully, we had not motored out far, so the beach was close. As my feet stumbled against the ocean floor, I half swam and half dragged myself out from the breakers. Joe arrived onshore a few moments later, and then the skiff washed up. We grabbed the bow line and tried to hold the skiff as the waves crashed down on us. But the sand gave way underneath, and it was as if a giant force, an undertow, pulled the skiff out to sea. We watched it go out half a mile.

I turned toward land. I was cold and knew I needed to get into dry clothes before hypothermia set in. With a sinking heart, I noticed we had washed ashore on the south side of the river. That meant we had to swim across the river to camp. I started to walk upstream and told Joe to follow. I wanted to find a section of the river narrow enough to swim across. Walking through the tundra tussocks helped keep me warm. I still had on my bright orange float coat and was grateful that regulations required us to wear them. Somewhere along the trek, I found the tip of a mammoth tusk and carried it in my hand for a while; I didn't notice when it slipped from my fingers. After a time, the river narrowed in width, and I decided I could swim across, so I kicked my boots off and, carrying them in one hand, plunged in, swimming a sidestroke. The river was slow-moving and not difficult to get across. I pulled myself onto the far bank, dripping wet yet again. Success! I would soon be dry and warm. Sitting to rest, I turned to see Joe still standing on the other side, waist-deep in the river.

"Swim across," I called to him.

"I can't swim," he replied.

"Dog paddle across. Your coat will keep you afloat."

"I'm not sure I can make it."

I could tell when anxiety freezes a person into inaction because that had once happened to me. In 1971, I became immobile halfway across a steep sheet of ice covering part of the Pacific Crest Trail I was hiking in the Sierra Nevada Mountains. I was fearful of slipping and falling onto the jagged boulders below. I knew that as time passed, Joe would get colder and less likely to attempt a river crossing to camp, his only means of survival for the next few days.

With resignation, I threw down my boots and plunged back into the river, swimming to where I had started. By the time I reached Joe, I knew what I would do. The training I had received in my Lifesaving class at the University of California Riverside came back vividly. If I could haul a football player across a pool in a test, I could carry Joe across the Tukruk River. My accomplishments in my undergraduate class gave me confidence I could succeed in this real-life test. I told him to lie quietly on his back. I would carry him with my right arm stretched across his chest and sidestroke with my left. He was to kick with his feet to help propel us across. For the third time, I swam across the river. My movements were slow and labored. I had lost all feeling in my limbs. It was by sheer willpower that I flung out my left arm. I stared at my left hand, willing it to come toward me in a stroke. Over and over, I repeated the messages in my mind—reach out and pull, reach out and pull. Two-thirds of the way across, I thought that my strength might give out. But no, I dug deep and found a spark of reserve energy. Crawling up the bank on the far side was a relief, but I had not yet secured my survival. I still had to get dry and warm.

Events became foggy as I slipped further into hypothermia. My cold-water immersion had lowered my body temperature and was causing mental confusion. I made it to my tent, although I don't remember how I got there. I remember stripping off my wet clothes, digging out some dry ones from my pack, and curling up into a ball in my sleeping bag like a small, burrowing animal. Then, violent shivering thermogenesis took over. I vaguely wondered if any of my bones would snap during the uncontrollable spasms that lasted about twenty minutes. I knew enough to be glad my body could still shiver; at least I wasn't too cold. It's when you can't shiver that you are dangerously hypothermic.

After the shivering stopped, I rooted around for food to restore my energy. My mind was working again. I saw that Joe was safe. We made a fire and settled in to wait for rescue. We wouldn't be overdue for four more days. However, a Cessna 185 on floats flew overhead in just two days. We jumped up and down, waving our arms and shouting, and the pilot waggled his wings to signal he had seen us. The plane circled and skimmed across the river until coming to a stop against the bank. Joe and I scrambled to throw gear into bags. I ran to the plane as the pilot opened his door and told him we were glad to see him, but our rescue was earlier than expected. He explained the skiff had washed ashore and been found by locals, who had radioed the troopers in Kotzebue. They sent a plane to look for survivors along the coast. That was us—survivors!

Although the skiff had lost its twin engines to the bottom of the Chukchi Sea, considering its ordeal, the hull only had a few dents and was still seaworthy. Outfitted with new outboard engines, Joe was at the helm again in a few days as a more experienced seaman. As for me, I was glad to be back in my Boston Whaler, sampling the coastline south along the Baldwin Peninsula. Sampling the waters off of Cape Krusenstern could wait.

As my research into the early life history of juvenile chum salmon came to a close, I concluded that chum salmon spawning areas in the Noatak River provide higher thermal exposure than experienced by chum salmon in more southerly parts of Alaska, suggesting chum salmon in the Arctic are not adapted to lower temperatures. Rather, they expanded their range north because of upwelling warm springs.

Juvenile chum salmon remained in nearshore waters until July and then moved offshore. I found no evidence of a food shortage or significant predation, so these factors would not limit additional fry production from a hatchery. The only limiting factor in the Kotzebue area seemed to be upwelling warm springs for spawning.[9,10]

I began the second phase of my job—investigating locations throughout the Kotzebue Sound area for potential chum salmon hatchery sites—by assembling historical reports of warm springs near villages. A key to site selection was abundant fresh warm water from upwelling springs. Another key was convenient access.

After identifying potential sites, I contacted the village councils. Before any site visit, it was paramount I explain to the village council the purpose of my trip and ask their permission to seek help in my endeavors, for which I would pay. When I arrived in a village, I first sought the village council office to introduce myself. A village elder listened to my quest and assigned a guide with a form of transport, such as a four-wheel all-terrain vehicle, to take me to the purported warm springs. After several site visits, we decided that my efforts to locate a suitable hatchery site should focus on the area surrounding the village of Noatak. Abundant warm springs were known to exist downstream from the village.

During my chats with Noatak residents over coffee, I heard about a warm spring seeping out of the ground, forming a small stream draining into the Noatak River. The place was located several miles below the village, on the western shore, and was called "Sikusuliaq," that roughly translated into English was, "place that does not freeze in winter." I added Sikusuliaq to my list of places to investigate as a possible hatchery site.

During the summer of 1980, I made several trips in my little skiff up the Noatak River, trying to find the stream that led to Sikusuliaq. One day I slipped into a small stream that was too shallow for my skiff, so I tied it off to a tree and followed the stream on foot to its source, which was a spring bubbling up from mossy tundra. As I carefully took measurements, hordes of mosquitoes and horseflies dive-bombed me, seeking blood. Thousands of them! Thankfully, I had a head net and rubber gloves, so the insects didn't have a chance.

After informing Jim of my find, he flew to Kotzebue and accompanied me back to Sikusuliaq. Pending detailed hydrological investigations, Sikusuliaq looked like a promising site for a hatchery. It was close to Noatak, so local people could be hired to staff the hatchery, a welcome economic benefit to the village.

ADFG operated the Sikusuliaq Springs Hatchery from 1981–1995 to augment wild chum salmon commercial and subsistence fisheries in Kotzebue. The hatchery was closed in 1995 due to decreased state funds. At peak production in 1992, the hatchery incubated 11,100,000 eggs. An estimated peak adult hatchery return of 90,000 chum salmon occurred in 1997, two years after the hatchery was shuttered.

Game Management and Research

My work as a Technician I for the Game Division took me to all the villages, where I was given a chair in the village council office to do business with the locals. I sold licenses, handed out permits, and sealed fur. Many people trapped lynx, fox, mink, marten, otter, wolf, and wolverine during winter and sold them as a source of income. Commercial fur buyers will not accept furs shipped to them without an official tag attached to the pelt. By documenting the species taken, sex, and approximate age, harvest records were kept up to date. Sealing fur also gave me an opportunity to interview trappers about their observations on animals since surveys on furbearers were not feasible.

My fieldwork on barren-ground caribou occurred primarily in the Kobuk River Valley. The Kobuk River, over 300 miles in length, originates in the Brooks Range and empties into Hotham Inlet. The drainage is vast and diverse, including glaciated mountains, steppe tundra plateaus, and the Kobuk Sand Dunes, a surprising 25 square-mile desert of shifting sands in the Arctic.

Thirty herds of caribou are in Alaska, distinguished by separate calving grounds. The state approaches their management by estimating herd abundance, subtracting annual natural mortality, and recommending harvest limits to the Board of Game based on production deemed surplus. If the herd decreases in size, the Board of Game will likely impose restrictions to prevent overharvest.

Dave and biologists from Fairbanks devised an ingenious way to capture caribou for radio collaring—lassoing them from boats as the caribou swam across the Kobuk River at Onion Portage. Lassoing caribou by their antlers was less risky and cheaper than darting them with immobilizing drugs, a common capture technique. Radio collars emit a signal that is picked up by a receiver carried by biologists during survey flights. Tracking radio-collared animals allows biologists to monitor movements and life history.

I was familiar with radio telemetry, having worked with desert bighorn sheep in the Santa Rosa Mountains during my years at the University of California, tracking their movements as an assistant on a National Geographic grant as well for my own research funded by a fellowship.[11] I had flown with biologists as they darted sheep with the immobilizing drug M99. Following examination, specimen collection, and radio collar attachment, the drug was reversed with M50-50. Darting stressed animals, so I hoped the lasso method would be less stressful on the caribou.

In the fall of 1979, I flew from Kotzebue to Ambler with two biologists from Fairbanks. A few days earlier, Dave and another biologist, Fred, had left Kotzebue by boat to meet us at Ambler. Ambler is a small village on the Kobuk River, about 140 miles northeast of Kotzebue. Ambler sits just upriver from Onion Portage, renowned as an archaeological site, with human occupation dating back 12,500 years. Onion Portage is the traditional site caribou choose to cross the Kobuk River as they migrate from summer grounds north of the Baird Mountains to winter grounds south of the Waring Mountains. We climbed into the boat, motored downriver to Onion Portage, and quietly waited for the caribou to come.

We hid behind bushes whose leaves had turned scarlet and gold from autumn. Toward late afternoon, I could hear the caribou approach the river by the clicking sound of their heels. Then, their hooves tapped upon riverbank gravel, and there was a soft splash as their bodies slid into the river. Their lightweight, hollow-shafted hair helped to keep them buoyant in the water as their legs paddled furiously against the current.

Suddenly, we jumped from behind the bushes, pushed the boat into the river, clambered in, and paddled after the caribou. At mid-river, we pulled out the lassos, and as the antlers were ensnared, the racks were hauled alongside the boat. Two people attempted to hold the head still by the antlers, which was not easy as the bulls plunged and struggled to free themselves. A third person dangled over the side of the boat, slipped the collar around the neck, and tried to attach the tiny screws and minuscule nuts, all without dropping themselves, the collar, or the hardware into the river!

Surveys to estimate the abundance and composition of caribou in the Western Arctic herd are conducted by ADFG biologists every year. In June and July, cows gather on calving grounds in the Utukok and Ketik River drainages east of Point Lay. During this time, biologists fly over the concentrated herd and count the number of cows and calves. Large groups of caribou are photographed with a camera mounted on the bottom of the plane. The photos are enlarged, and individual caribou are counted. Spotters on the ground overlook the caribou with telescopes to note the sex and age of thousands of caribou. The percentages of cows, calves, and yearlings seen by the spotters are multiplied by the total number of caribou counted by aerial surveys to give an estimated herd composition at the time of calving.

Additional surveys are conducted in October to count bulls and in winter to determine distribution and survival. Monitoring the herd during winter is important because snow conditions affect the ability of caribou to reach the grasses and lichens they need. If they can't get to the food, some will starve, and the herd will decline.

In April 1980, I put on warm clothes and, along with a fellow biologist, got into a Cessna 185 on skis bound for the caribou winter range. My job was to record the number of calves and cows to estimate the calf-to-cow ratio. When compared to last July's ratio, we would get an estimate of overwinter calf survival, a significant herd limiting factor. We spotted a large group of caribou, so the plane landed and skidded to a stop, depositing us onto the hard-packed snow some distance from the herd. It was a bright spring day, but the wind had a bite to it.

We quietly set up our telescopes and watched the animals as they wandered across the horizon. Caribou don't stand still; they are always moving on to the next tuft of something to eat. Some travel up to 50 miles a day. For several hours, I recorded the composition of the herd as they migrated through my field of vision. The only sound was the faint tinkle of tiny ice crystals as the wind puffed them across the frozen ground. It was cold, stiff work, but it was enjoyable to watch the animals.

Caribou walking along a snow-covered plateau

A spotter notes sex and age of caribou in April 1980

In addition to village visits and fieldwork, Dave asked me to write public relations articles for the local newspaper, the *Mauneluk Report*. Unofficially, I became acting Area Game Biologist for three months in 1979 while Dave went on leave. I was never credited with this service nor given the salary that went along with my designation. That's just how things were done back then.

Village Life

The wind always blew on the coast. I was told a story about the wind that went something like this. On a winter day long ago, a child wandered onto the sound and was tumbled by strong winds for miles across the ice. When searchers finally found him, he was shaken but fine because he had stayed warm, dressed in the traditional fur parka and mukluks. As the streets became icy one windy day, I tested this story. I turned my back to the wind and held my arms out to catch more air. The wind slid me down the road on my bunny boots! The story was plausible.

As the snow deepened, I decided to get a Bombardier Ski-Doo 440 and sled for transport and winter recreation. Just as I had learned to "read" water for safe boating, I also learned how to read signs in the snow and ice for safe snow travel. My first encounter with overflow was on Kotzebue Sound. Overflow can occur on sea ice or river ice. On sea ice, water can seep up through cracks from tidal action and become covered by snow. It's hard to see this slush before your snowmobile suddenly mires and sinks. I was lucky because I was only mired in 6 inches of overflow. The danger is getting wet while you pull the 440-pound snowmobile back toward solid ice. Wet clothes freeze and become stiff, making it harder to move and eliminating insulation against cold. After my first encounter with overflow, I carried a pole in the sled. If I had doubts about the surface ahead, I stopped and gingerly walked forward, testing the snow with my pole. Although it's slow going to test the route for overflow, it's easier and faster than getting stuck.

Another sign I learned to look for while traversing the frozen Kotzebue Sound was sea ice ridges. The sea ice shifts and cracks, creating ridges hidden under the new snow. It is quite jarring to be racing along only to have your snowmobile runners jammed to a sudden stop or be flipped over sideways. Occasionally, warm Chinook winds open up leads in the sound. At these times, travel on the sound is treacherous.

Winter is a time for skin sewing, and I approached this newly discovered craft with zeal. I cut and sewed a deer hide into mittens, bought river otter fur to sew on as trim, and sewed on beads for decoration. A pair of heavy wool liners completed the ensemble. I bought a pair of mukluks made from the hide of caribou legs, with soles from the tough skin of oogruk. The seal skin was water-resistant and sturdy. I trimmed my new mukluks with a matching bead pattern. I enjoyed wearing my mittens and mukluks because they were warm and helped me feel closer to the Iñupiat culture.

Iñupiat skin sewing finery was on its best display during the Arctic Circle Championship dog sled races held on Kotzebue Sound on April 18, 1980. At -2° F, the temperature was just right for the dogs that grinned and yelped in excitement. People lined up on Front Street, showing off their parkas made from spotted seal or ground squirrel skins trimmed with wolverine tassels. The ruffed hood was usually wolf, lined with wolverine to keep the wind from biting your face. Wolverine fur has tapered tips so

Fur parkas

the frost can be brushed off easily, unlike fox fur, which holds your breath in little ice balls. The price of a wolf ruff was $400, while an Arctic fox ruff sold for $50.

As spring approached, the blessed sun reappeared. My first Arctic winter left me starved for sunlight! If the weather was tolerable, I joined friends for a weekend of cross-country skiing. They had a canvas tent and wood stove set up at a site in the lower Noatak River valley. I loved skiing through the trees in soft snow, watching for tracks of Arctic hares, snowshoe hares, and foxes.

A novel gardening experiment was organized in Kotzebue by Tony Schuerch. Tony was convinced that traditional fishers and hunters would benefit by learning about crop cultivation. The cost of flying in produce made fresh vegetables expensive at the local KC store. The Mormon Church in Kotzebue donated land for garden plots, and I enthusiastically signed up, along with a few other determined gardeners. The constant coastal winds buffeted the small plants all summer. By September, I had tiny potatoes, short carrots, and a harvest of Swiss chard.

To make up for my skimpy vegetable harvest, I turned to berry-picking in the fall. After the first frost, I joined scores of women and children scouring the tundra for blueberries, crowberries, low-bush cranberries, and salmon berries. I walked miles across the gold and red-leafed hills above Kotzebue—it was a berry picker's delight, especially since the swarms of mosquitoes and gnats had died down. I made the blueberries into jam and strained the seedy crowberries to make a jelly that tasted like grapes.

I joined a food cooperative and ordered supplies in bulk that arrived by barge during the two-month window of open water. A family needed to order a year's worth of dry goods, such as flour, powdered milk, and dried beans, to last until next year's barge arrived. Subsistence harvest of fish and game supplemented my food supply. To catch chum salmon, friends and I bought an old gillnet, 75 feet long. I repaired the holes by winding braided twine around a shuttle, then wove the shuttle through the ends, tying knots to make a 6-inch-square mesh to target chum. The net has to hang just right, and the mesh must be of the correct size to tangle a fish by its gills. If the mesh is too big or the net has holes, fish will swim through the net. If the mesh is too small, they'll bounce their nose off the net and swim around it.

Being newcomers, we selected a fishing site no one else wanted south of Kotzebue, where chum salmon are not likely to migrate. Native families in Kotzebue laid traditional claim to the best sites for netting salmon, with families fishing in the same place for decades. I took my turn checking the net at regular intervals and was thrilled to find my first two salmon flopping in the net. We cut the fish into strips, hung them on a rack to dry in the sun, then lightly smoked them—smoke is a traditional preservation method because it slows decomposition.

For fresh meat in winter, I traveled by snowmobile across the frozen sound and into the lower Noatak River hills, searching for snowshoe hare tracks among the willows. After spotting tracks, I patiently crouched behind the snow-go and had my .22 rifle ready for a white hare darting through the thicket. A triumphant outcome was a couple of hares for the crockpot. The best way to cook a sinewy hare was to make a stew.

My first big game hunt occurred in October 1980. Two friends who worked for Commercial Fisheries and I chartered a plane to land us on the gravel bar of a small river southeast of Kotzebue. We set up camp next to the river. The snow-dusted hills above the river valley were beautiful, and it was mostly sunny weather, with a little snow flurry now and then. The next morning, we got up early and distributed ourselves in the woods around the edge of a meadow. I'd learned to make the grunting sounds of a female moose to lure in big bulls to mate and proceeded to "call" in bulls. To my utter shock, it worked, and a big bull came crashing through the brush toward me. He startled me half to death! Dan shot at him with his .223 rifle, which I thought was an underpowered gun for the job. We could see that the bullet had wounded the bull because we spotted blood on a blueberry bush.

We tracked the bull upriver, where I dropped him at the water's edge with one shot from my Remington .30-06. The hunt was over by 11:30 a.m., but we had lots of work to do. It took us two and a half days to cut up and pack all the meat back to camp. There was one exciting moment when a grizzly bear burst onto the scene, and all Joe had to ward off the bear was an ax. Miraculously, swinging the ax at the bear convinced him to back off.

To hasten the process of getting meat back to camp, Joe lashed tree limbs together into a small raft. We piled on the remaining chunks of meat and the antlers, tied ropes on the raft, and lined it downriver to camp. At night, a wolf pack howled in the hills above camp. In that wild and primitive landscape, I felt a hint of what early humans must have experienced living in the Arctic. Back in Kotzebue, we divided the moose into equal shares and took a week to wrap the meat. Every bite tasted good because I had worked hard for it. The guys thought I deserved the antlers, which were 64 inches across, a big bull!

I had always planned to be a transient of the Arctic—to enjoy adventures, learn about the ecology and culture, and use my experiences in Kotzebue as a springboard to a better-paying and more secure job. A permanent position! My assignment with FRED was ending, and the time had come to begin applying for other biologist positions. I felt unsettled; I wanted to know what I would be doing in the coming years.

The Bureau of Land Management was hiring a wildlife biologist for a reindeer/caribou range interaction study. I applied and flew to Nome for an interview, but they hired someone else. I applied for several permanent Game Biologist positions at ADFG that were opening in various areas. While interviewing for these positions, I was stunned to discover the challenges awaiting a single woman biologist on the Last Frontier. For the assistant area game management position in Dillingham, I was told the town, a commercial fishing hub of Bristol Bay, was too rough and rowdy for a single female, and supervisors did not want to be responsible for my safety. Cross that job off the list. In another interview, a biologist candidly told me his wife would kill him if he hired a single female as his assistant. The job entailed periodic overnight travel to outlying areas, and he didn't want to deal with the flak from his wife. With chagrin and a sigh, I told him I appreciated his honesty. Cross that job off the list. In my final interview, supervisors told me they didn't think male hunters and trappers would take a female game biologist seriously. How does one argue with that perception on the part of stakeholders *or* supervisors?

In contrast to the constraints offered by the Game Division in 1980, the Commercial Fisheries Division was delighted to hear from me. To be honest, they were interested in just about anyone breathing and standing upright—the need was great, and there were few qualified takers for hard work in often nasty weather at remote locations. A biologist I had met in Kotzebue suggested I apply for a permanent Fishery Biologist II position opening up with Commercial Fisheries Division in Glennallen. The job was to supervise a sonar station counting Chinook and sockeye salmon returning to the Copper River drainage, central to managing those commercial fisheries. The supervisor, Ken, interviewed me over the phone. Then, I waited. My part-time nonpermanent Fishery Biologist II position had been cut, and I was at the last paycheck.

While I was job hunting, I gave thanks for what I had (shelter, a full refrigerator, the ability to earn a living) and appreciated the simple needs that make life pleasant. By January 1981, I was packing to leave Kotzebue. I had been hired for the permanent Fishery Biologist II position in Glennallen! I was a bit anxious, but my new boss made me feel so welcome that I relaxed. Everyone said it was a good career move. I left Kotzebue with wistful feelings. I felt I had been transformed by all I had learned from the Arctic and its people, but a new beginning meant meeting new people and more adventures.

I flew to Anchorage, got my car and belongings out of storage, and drove 180 miles on the Glenn Highway to Glennallen. The drive took about three and a half hours.

Copper River Salmon

"You cannot get through a single day without having an impact on the world around you. What you do makes a difference, and you have to decide what kind of a difference you want to make." ~Jane Goodall

The high volcanic peaks of Mt. Drum and Mt Sanford, with Mt. Wrangell in the background, commanded the horizon as I drove my Ford Bronco into the small community of Glennallen on February 1, 1981. With a population of 511 in the 1980 U.S. census, Glennallen sits at the crossroads of the Glenn and Richardson highways in the Copper River Valley. The valley is huge, encompassing 20,650 square miles of primarily dwarf black spruce, white spruce, and aspen forests dotted with small, scattered settlements whose inhabitants totaled around 2,500. The Ahtna Athabascan people were the valley's first residents. In the late 1800s, prospectors built trails through the valley to connect Interior gold mining communities with the Port of Valdez. During World War II, the trails were widened into roads.

The pulse of the valley is the Copper River, which originates at the base of Mount Wrangell and flows for about 300 miles through the Wrangell and Chugach mountains to Prince William Sound, draining mostly pristine wilderness. The river's gray silt-laden waters, fed by glaciers, roil down a steep grade at high velocity. In summer, rainfall, melting snow, and glaciers contribute significant flow to the river, averaging 113,000 cubic feet per second.[12] Abundant runs of sockeye and Chinook salmon, which migrate up the river's myriad tributaries to spawn, support a thriving commercial fishing community in Cordova, and nurture upriver sport, personal use, and subsistence fishers as well.

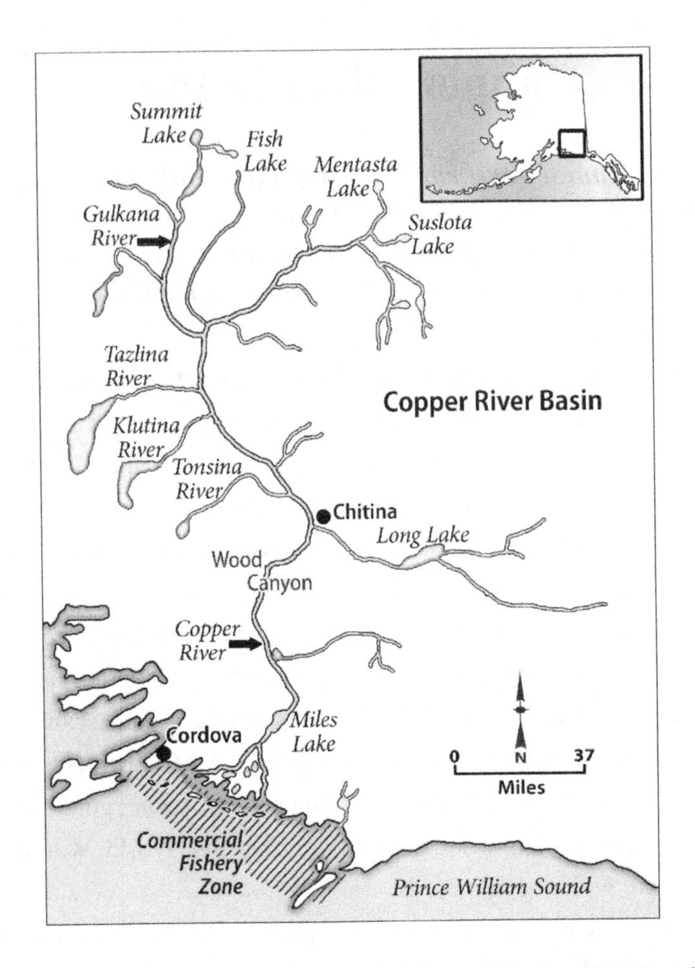

I arrived in Glennallen less than two months after President Carter signed a bill creating the Wrangell-St Elias National Park and Preserve. As in Kotzebue, there was unease in the local populace about how federal jurisdiction would impact people's interactions with the land and river.

My job was to continue the study of salmon in the Copper River, which had begun in 1967, with the aim to manage sustainable harvests through escapement goals. Initially, it was impossible to count migrating salmon in the glacier-fed river because they couldn't be seen through the silt. This problem was solved in 1978 with the installation of Bendix side-scanning sonar, expanded in 1979 to two counters on each bank of the river at the Miles Lake camp, about 33 miles upriver from the commercial fishery district in Prince William Sound. Sonar counts provided daily estimates used by managers to ensure an adequate escapement of sockeye salmon into the upper Copper River while allowing for sustainable harvest.

My responsibilities were to run the sonar project out of Cordova from mid-May to early August and to conduct aerial survey counts of salmon spawning in clear water tributaries in August through September, with flights based out of Glennallen. I frequently traveled between these two bases of operation, Cordova and Glennallen, over the next three years. During my stay in the Copper River Valley, from February 1981 through March 1984, I embraced the pioneer spirit of a self-sufficient community and watched as they argued with and adapted to the inevitable changes revolving around them.

A white bungalow on the south side of the Glenn Highway housed the troopers in one half and ADFG in the other half. I parked my Bronco behind the building, climbed old wooden steps, opened the door, and walked into my new office home. The fragrance of fruit cake added to the warmth of the place. I learned that Kathy, the secretary, often treated biologists to her homemade cakes. I introduced myself to Kathy and said I was looking for my new boss, Ken. Kathy stood up and, with a smile, motioned me down the corridor. As I stepped inside Ken's office, he looked up and exclaimed, "You're not a blonde!?"

Later, we laughed at his first words to me, but this odd greeting puzzled me at the time. Ken explained that a biologist in Kotzebue had highly recommended me for the job but had described me as blonde; a joke on Ken, I suppose. Ken rescued the awkward situation with a warm welcome and a tour of the office.

Commercial Fisheries Division had two biologists, Ken and me. Russ, who was assigned to a sockeye salmon enhancement project, worked for FRED. Sport Fish Division had two biologists, Fred and Butch. The Game Division had an area manager, Bob, as well as a wolf and bear research component supervised by Warren and aided by his assistant, Jack. In the summer, seasonal technicians added to the crowded nature of the small office. I soon became acquainted with everyone and enjoyed their company. Kathy and Bob liked to run along the shoulder of the Glenn Highway during the lunch hour, and I joined them, working up to 3 miles a day, weather permitting. I found myself more spirited and cheerful after a run through the brisk winter air, although one day, I came back in with frost-nipped ears.

Biologists in our close-knit office pitched in and helped each other when projects were temporarily short-handed. The variety of fish and game work expanded our view of the resources in the valley and the people who depended on them. The sockeye salmon enhancement project was conceived by Ken when in 1971, he discovered warm springs in the upper East Fork of the Gulkana River, a tributary to the Copper River. The upper East Fork is an important spawning area that significantly contributes to the production of sockeye salmon in the Cooper River. The springs provide a stable growth environment where salmon eggs incubate over the winter. Previous years of low sockeye salmon escapement and subsequent fishery closures had highlighted the need to increase fish production to meet user demand.

By 1973, a low-cost streamside incubation pilot project was launched and proved successful. Rows of boxes filled with gravel were placed in the stream, so the water flowing through carried oxygen in and waste out.[13] In the fall of 1981, I helped staff collect ripe female sockeye salmon and extrude and fertilize eggs with milt squeezed from males. With utmost care, eggs and milt were collected in a bucket and gently mixed. Eggs were treated to prevent fungus development. After an hour of water "hardening," the fertilized eggs were carefully placed on gravel in the boxes. Eggs protected and monitored in the boxes have a much higher rate of survival to the fry stage compared to eggs naturally strewn in redds by females, which are subject to predation and unfavorable environmental variability.

Because of my experience in the Palmer Game Division office, I helped Bob conduct aerial surveys of moose and caribou herds in GMUs 11 and 13 a few times. The Mentasta caribou herd ranged primarily on the northern and western slopes of the Mentasta and Wrangell Mountains. The herd was small but in the early 1980s, was thought to have sufficient productivity to withstand limited harvest. I remember sitting in the passenger seat of the super cub as the pilot wrestled with the force of the winds uplifting from Mt. Sanford, when a sudden gust bounced us so much that my head hit the ceiling of the plane hard—ouch!

The wolf research project monitored wolf movement with radiotelemetry and studied their population dynamics, focusing on the impacts of predation on moose and caribou. A better understanding of wolves would help shape decisions regarding wolf management policy in the area, specifically addressing the controversy of wolf control measures.

One day, I went with Warren and Jack to dart and radio collar wolves from a helicopter. My job was to help take and label samples from the immobilized wolf before the drugs wore off. In the early afternoon, I was kneeling in the snow, the open box of sampling equipment to my side and a female wolf lying in front of me when she suddenly came alert, rose to her feet, and looked level into my eyes, which were about 20 inches from her eyes. I was mesmerized, immobile, like a cobra's victim. Her amber eyes were fearless and forthright, their depth revealing the primeval history of her species. After a few breathless seconds, she turned from me and calmly trotted off into the dwarf spruce, quickly disappearing. I've always believed that her choice to turn from me was because we had somehow communicated an acceptance of each other's presence.

Me and the wolf [14]

With snow thick on the ground, there was no Commercial Fisheries fieldwork for several months, so I settled into a corner desk at the office and attended to writing overdue reports on sonar studies in the Copper River.[15] I liked to write. Most of the biologists I had met did not. I analyzed the data Ken had collected during the previous two years with a small hand calculator. There were no computers in the area offices in those days. Only the biometrician in the Anchorage office had a computer. I had to make an appointment and drive 180 miles into Anchorage if I wanted statistical procedures run. Thus, producing a report took some time.

In addition to the overdue sonar reports, Ken gave me a heavy and intriguing file of tag and recovery data collected between 1967 and 1972. Before sonar was installed, biologists used fish wheels to capture, tag, and release sockeye salmon at Wood Canyon. The date of capture for each fish was noted. Tag recovery occurred in the upriver subsistence fishery and at fifteen tributary streams. This massive research effort resulted in the tagging of 87,823 sockeye salmon across six years. The project was designed to estimate escapement, but what had gone unnoticed was that the data could also be used to examine migratory behavior. Salmon stocks maintain a chronological order during their spawning migration. By understanding when different stocks of varying productivities move through a harvest area during a given time frame, managers can schedule fishery openings and closures throughout the season to distribute harvest across groups of stocks equitably so no stock is over-harvested.

In my spare time over the next several years, I plowed through this data archive that eventually revealed the migratory timing of stocks through the lower river. Knowing the mean date of migration through the Copper River commercial fishery district allowed managers to assess the probable impact of harvest on groups of stocks. In 1986, the completed analysis of this data archive resulted in a paper that was published in a professional journal.[16] I was gratified to have helped produce an aid for salmon management.

As the field season approached, I held interviews to hire five field technicians whom I would supervise, and attended staff meetings in Anchorage and Cordova to review pre-season forecasts, management strategies, and research projects. The increase in activity heightened my anticipation about running a sonar camp.

A week before I was to leave for Miles Lake camp, sonar project leaders around Alaska were summoned to Soldotna for an intense three-day workshop. Our group learned how to assemble and deploy sonar substrate on the Kenai River, hook up electronics, and sight on the target (salmon swimming upstream). In the evenings, we discussed hydro-acoustic theory, learned how to calibrate the sonar count with visual interpretations of fish passage on the oscilloscope, and reviewed sonar data collected on different rivers. At the end of the workshop, I felt confident and prepared for supervision of the Miles Lake camp. But, the greatest challenge to running the Miles Lake camp was the Copper River—no workshop could have prepared me for what I was about to encounter!

The Miles Lake camp sits at the end of the Copper River Highway, 49 miles inland from Cordova, a beautiful small fishing town near the mouth of the Copper River. Home to the Eyak and Chugach peoples, the area was also visited by Ahtna Athabascan and Tlingit peoples. In 1790, the area was discovered by Spanish explorer Salvador Fidalgo who named it "Puerto Cordova." The town materialized in 1906 when it became a port for the Copper River and Northwestern Railway built by J. P. Morgan and the Guggenheim family to transport copper ore from the Kennecott mine in the Wrangell Mountains, 196 miles from Cordova.

The town sits in a temperate rainforest with cool temperatures and heavy rainfall. Raindrops splattering the truck's windshield were as big as my fist. The average annual rainfall is 167 inches, and snowfall is 80 inches. During my first season, it rained continuously for thirty-five days. We always kept our raincoats handy.

One of the first things I noticed about Cordova was its bustle as a busy fishing town. In the 1980 U.S. census, there were 1,879 inhabitants of Cordova. Hatcheries that augmented chum and pink salmon fisheries contributed to the town's brisk nature. Private non-profit and state-run hatcheries influenced the culture and economy. During 1981 and 1982, the hatcheries incubated from 170 to 180 million salmon eggs for release as fry into Prince William Sound.[17] By 1983, additional hatcheries had come online, bringing total capacity to an astounding 225 million eggs.[18] Prodigious adult returns from released fry in the coming years kept managers in overdrive monitoring fishing openings. and the purse seine fleet scrambling to harvest millions of fish. Local processors ran at maximum capacity.

I was given a state truck and a small office in the Cordova ADFG building that I shared with Ken on his weekly visits from Glennallen. I drove from camp into town a few times a week to pick up mail, buy groceries and work in the office. Communicating closely with Rich, the area manager, was a good idea. Rich explained that the Prince William Sound area is divided into eleven commercial fishing management districts, the most prominent being the Copper River District, which my sonar project addressed. This district supports the drift gillnet salmon fleet. All five species of salmon are harvested here, although the district's target during spring and summer is sockeye salmon. The district fishery harvests nearly all of the area's Chinook salmon and about 65 percent of the sockeye salmon. Commercial fishing begins mid-May and is regulated by field announcement. The sockeye salmon fishery is completed by the end of July, with most of the catch taken between late May and June.

ADFG had a bunkhouse in town, and occasionally I stayed overnight for staff meetings or to attend advisory committee meetings. Advisory committees are a crucial component of Alaska's citizen-based fish and game management and allocation system. The committees, located throughout Alaska, are made up of stakeholders who submit proposals to the Boards of Fisheries and Game for regulatory changes they believe are appropriate for the resources and users in their area.

On May 10, we loaded the truck with our gear and supplies and set out from Cordova to the sonar camp on the Copper River Highway, built on the abandoned railbed of the Copper River and Northwestern Railway. I was eager to catch my first glimpse of the sonar camp, where the crew and I would spend the summer. Ken drove, so I gazed at the scenery, intent on taking in all the details of this new and exciting journey. As we progressed inland toward the massive glaciers framing the Copper River, the temperature dropped, and the advance of spring reversed. Full-leafed trees gave way to bare branches, green faded into gray rock, and emerging life receded into dormancy. I began to feel the effects of the changing landscape, which left me subdued as I reviewed my grave responsibilities in this desolate wilderness.

Snow drifts blocked the remaining 10 miles of the road to camp. Ken was prepared for snow drifts—we had hauled a snow track vehicle behind the truck for the last leg of the journey. My first sight of the camp was a massive snow drift that half-buried the buildings.

Beyond the camp, the Million Dollar Bridge loomed, black steel girders stark against gray mountains. In 1906, engineers with the Copper River and Northwestern Railroad decided a narrowing of the Copper River, which occurs just below Miles Lake, was a logical site for a bridge. Named the Million Dollar Bridge for its $1.4 million cost, it was built under great difficulty, and several lives were lost. It spans the river between Miles Glacier, above, and Childs Glacier, below the bridge. Massive concrete icebreakers were installed on the upriver side of the 1,550-foot bridge to protect the piers from icebergs sent downriver from Miles Glacier.[19] The highway was supposed to cross the Copper River at the Million Dollar Bridge, but the 1964 Good Friday earthquake knocked the north span of the bridge into the river, abruptly terminating the highway. The bridge was left unrepaired for decades, leaving a highway to nowhere.

The Million Dollar Bridge across the Copper River, 1981

We had a lot to do. Snow shoveling was necessary to reach the buildings. The main cabin was for cooking and eating; this was where the radio was housed to call in daily sonar counts to the Cordova office. I slept in the loft of the main cabin, accessed by a ladder through a trap door. There were two smaller cabins as crew quarters. The camp, built just three years previously, was still being developed, and more construction lay ahead of us.

After putting gear and supplies in the cabins, Ken explained that typically in early May, glacier melt and rainwater have not yet swollen the Copper River, so the water level is low, thus exposing the south bank. As the sonar site was being developed in 1978, the exposed south bank presented the opportunity to construct a permanent substrate of concrete that rests on the river's bottom, perpendicular to the river's flow. Ken set abandoned rails from the old railroad in concrete for a track to slide the transducer along. The transducer is a box that emits the sonar. The permanent substrate is used during high water. A temporary substrate, made of three cylindrical aluminum sections bolted in tandem, is deployed during low water.

Main cabin, Miles Lake sonar camp, 1981

At the start of the season, we used the temporary substrate. As summer progressed and the water level rose, we transferred the transducer to the permanent substrate. Because the Copper River fluctuated several feet vertically per day and 15 feet or more during the counting season, we went back and forth between using the temporary and permanent substrates.

My first task as Project Leader was to learn how to deploy the temporary substrate. The aluminum sections were retrieved from behind the main cabin, hauled to the water's edge, and bolted together. Heavy wire cables were anchored onshore and affixed to the 60-foot-long cylinder. The next part was tricky and dangerous. We eased the cylinder into the river and let the current turn it perpendicular to the shore, then used the come along to fine-tune its position. The cable quivered from the tremendous force of the river as it held the cylinder taut. Had the cable snapped, it could have broken our legs. Our relief at a job well done was short-lived. Icebergs hurtled downriver from Miles Glacier and collided with the cylinder, turning it into crumbled aluminum foil.

The temporary substrate lays on the south bank, crumbled by an iceberg. Child's Glacier is in the background.

Thankfully, we had spare sections for situations like this. We waited a few days for the river to clear of ice before we deployed the newly patched temporary substrate again. We hooked the sonar counter and oscilloscope to the 12-volt battery and installed the solar panel on the south bank shed that housed the electronics.

I gave a basic overview of the sonar equipment to the crew. The submerged transducer emits a beam of sonar, which is soundwaves. The beam returns echoes from fish that are picked up by a counter. The beam is aimed low enough so fish will not swim below it but high enough so river bottom debris will not disrupt the detection of fish with "acoustical noise." The counter automatically prints fish counts on paper tape. Counts are verified by a person who views the echoes on an oscilloscope.[20] After I showed the crew how to operate the equipment, I set up a schedule to watch the oscilloscope trace of the beam every six hours, for a total of four hours per day, to compare visual data with sonar counts on both banks.

A salmon's decision to swim upstream and how fast it swims is influenced by water level, light, fish numbers, and whimsy. Sometimes, when I watched the oscilloscope, I'd see a fish linger in the beam, slowly drift downstream, only to swim back into the beam, hold there for a while, drift backward again, and finally swim upstream. This was one fish, but each time it entered the beam, the counter ticked it off as a new fish, giving a false count of three fish. Another time I watched the oscilloscope green blips exploded on the screen. I poked my head out of the shed to look down on the river and discovered a group of river runners had arrived and pulled their kayaks out on the south bank. Other events that impacted the counts included rainstorms that produced a heavy vegetation debris load in the river and of course, icebergs.

The crew and I were tasked with additional camp construction work in between chores associated with keeping the south and north bank sonar counters running. One such project was to extend the permanent substrate on the south bank farther down the bank while the river was low. This meant mixing and pouring concrete.

A new technician was an eighteen-year-old teenager named Charlie who decided he wouldn't take orders from any woman who didn't carry a bag of cement. I patiently informed Charlie that I wasn't hired to carry bags of cement—he was. However, if it meant that much to him, I'd strike a bargain. If I carried a bag of cement down the bank to the construction site, then he would do as he was told and do a good job of it. We shook hands on the deal. I went to the truck and hefted a 100-pound bag of cement onto my shoulder. Given I weighed 125 pounds, the only way I carried that bag was due to sheer stubbornness. As I tossed the bag at Charlie's feet and stood looking at him, a flicker of acquiescence shone in his eyes. From then on, he was a good employee. I encouraged him to go to college and wrote a letter of recommendation for him. That fall, he was accepted into the University of Alaska Fairbanks biology program and eventually became an area manager with ADFG. I was proud of him and glad to have played a positive role in his life.

The 1981 sockeye salmon commercial fishery opened on May 18 with 392 boats in the water. As the strength of the run became apparent through sonar counts and catch trends, fishing periods were adjusted to allow for the desired upriver escapement of 250,000 to 350,000 salmon established by the Board of Fisheries. I felt my work was important to the economy and well-being of communities that depended on Copper River salmon.

Local residents and curious tourists showed up at the camp often, interested in tours of the facility. On average, I gave tours to six groups per week. I knew how important it was to inform and educate about the sonar program and to address concerns so the public would have confidence in management of the fishery. Our equipment was subject to periodic vandalism despite our efforts to garner goodwill.

In early June, Al Menin, the engineer who had developed the sonar for Bendix Corporation, came to stay for a few days to check the electronics. It was his habit to travel to rivers where his sonar counters were in use and provide annual oversight and maintenance. Al was a diminutive, cigar-chomping man who loved his work and enjoyed visiting with biologists at the various sonar camps throughout Alaska. Because the Copper River was running a small debris load, he turned up the counter sensitivity.

One sunny day a black bear walked into the middle of the camp, likely drawn by curiosity as well as a whiff of food. We yelled, banged pots, and made all kinds of noise to scare him off, but he wouldn't budge. I got the shotgun from the cabin, loaded two shells, and fired them into the air, but he was not impressed. I shot at his feet to scatter dirt and rocks on him, and he moved a fraction. I didn't want to kill him if I could avoid it. I kept loading shells and firing at the rocks near his feet, hoping to convince him to move off. After several rounds of this, he finally turned and slowly ambled off toward the riverbank across from Child's Glacier, where he knew he could eat some rotting fish carcasses in peace. I was glad to see him go and also glad he never returned to our camp.

It was interesting how salmon carcasses ended up on the riverbank across from Child's Glacier. Every time the glacier calved, the force of the iceberg as it hit the water created a large wave that crashed onto the opposite riverbank, depositing everything passing on shore. If salmon happened to be swimming by at that moment, they were flung into the air and plunked on the bank. We could always tell when a calving event occurred because it shook the cabins like an earthquake. If this happened at a convenient time, we ran down to the beach to pick up a few fresh fish for that night's dinner.

While workdays were long, I had some free time to jog down the highway and enjoy the spectacular scenery. Around camp, I saw moose, deer, bears, rabbits, geese, ducks, swans, and bald eagles. When I told the game biologist in Cordova that I had seen mountain goats across from camp, he gave me a spotting scope and asked if I would record my observations. I was happy to oblige. On June 13, my journal entry read, “I saw five adult mountain goats and a newborn kid.”

On July 6, some of the crew and I motored upriver in the skiff to explore Miles Glacier. As we neared, I beheld a massive frozen otherworld, sparkling and deceptively motionless. Dark metallic mountains flanked white seracs poised to tumble. It was a breathless day; the winds were still. We dared not take the boat closer and were content to explore the nearby riverbank. I saw a speck of green. It was a little tree, all by itself, struggling to survive. To me, the tree symbolized the miracle and resilience of life in the face of daunting adversity and brought a smile of wonder.

A lone tree grows on the Miles Glacier moraine

High sonar counts and salmon catches revealed that 1981 returns were stronger than anticipated, so everybody was happy. We had already achieved 75 percent of our desired escapement by June 9. On June 23, we surpassed our escapement goal. An entry in my journal read, “I am enjoying my summer. My boss and I get along. ADFG has a family feeling, camaraderie.” I think the feeling of "family" grew from our reliance on each other in working toward a common purpose under often difficult conditions. We depended on each other, and we had each other's backs.

The summer of 1981 at Miles Lake was unusually wet. On July 19, my journal read, "We've had rain for thirty-five days straight." The crew became depressed, with a few cranky souls, no doubt stressed by working in a continuous downpour. The Copper River got so high that a bridge to town washed out, and we were stranded at camp. My superiors devised a plan to bring us supplies via floatplane, and radioed the plan's details to me.

At the appointed time, I heard the plane as the pilot circled above camp to notify us that he was ready to land in the Copper River. Three of us scrambled down the bank to get into the skiff. I saw the plane glide onto the roiling water above the Million Dollar Bridge, with the nose of the plane pointed upriver as we dashed across from the south bank to rendezvous with the aircraft. I carefully steered the skiff to come alongside the nearest float, and my crew grabbed ahold of the strut. The current carried us in tandem downstream toward the bridge. We had less than a minute to catch boxes and bags the pilot threw out of the back of the plane into our skiff. All the while, I watched over my shoulder to gauge how close we were to the bridge pilings. At a critical distance, we had to let go of the strut, regardless of whether we had secured all the supplies because the pilot needed to restart the engine before the current carried the plane into the bridge. As I skirted the skiff around the concrete icebreakers, I held my breath as I waited to hear the engine come to life and see the plane plow a wake upstream in its advance skyward. He made it! The bridge to town was repaired on August 1. While dicey, the improvised supply plan had been pulled off without any mishaps.

Sonar counts ended on August 9. We closed down the camp and relocated to Glennallen. The total count of 534,263 fish exceeded optimum target levels. The total commercial harvest of 487,000 sockeye salmon was above the 10-year average, and the catch of 20,782 Chinook salmon was also deemed good. The estimated value of the 1981 sockeye salmon harvest in the Copper River District (based on the price per pound paid by local processors to fishers) was $4.64 million, and of the Chinook salmon harvest was nearly $900,000.[17]

Based out of the Glennallen office, salmon enumeration in the Copper River transitioned to aerial surveys. Counts of salmon in a group of streams, deemed representative of escapement throughout the drainage and with reasonable visibility, are surveyed yearly to provide trends in the number and distribution of fish on their spawning grounds.

Ken accompanied me on my first few aerial salmon surveys and trained me in the proper procedures for counting and recording observations. I learned aerial survey estimates of salmon are influenced by timing, so index streams are flown multiple times to ensure the peak of that stream's escapement has been observed. Flying in a slow and low-flying Super Cub or a Cessna 185, we could count small numbers of salmon, but the numbers of large groups of fish had to be estimated. I loved glimpsing the cheerful colors of red salmon in shallow, blue-tinted streams lined with the dark green of spruce and yellow hints of aspen foliage, all dwarfed by the imposing cold-white glaciers of the Wrangell Mountains. Small and remote tributaries of the upper Copper River country have no access, so I valued the opportunity to see and work in this isolated, striking country.

In 1981, the aerial survey count of twenty index streams totaled 76,820 sockeye salmon, the highest since surveys were initiated.

While the abundance of spawners upriver was high, some streams had too many, and others had none. Uneven distribution of spawning salmon suggested the stocks were subjected to disproportionate harvest levels in the commercial fishery, where some stocks were over- and others under-harvested.

In September 1981, I completed my probation period and was officially on permanent status—I had waited for that for a long time! At my hire as a permanent employee, I had close to three and a half years of temporary time. In 1978, the legislature altered criteria for state employees, stating that any time earned as a temporary could be applied toward total years' service. However, the time would not count toward vesting.

In October, all newly-hired permanent biologists were directed to attend the Alaska State Trooper Academy in Sitka for a week of training. Following completion of instruction, we would be deputized to enforce Alaska Statute Title 16 Fish and Game laws to complement enforcement efforts by troopers in the Fish and Wildlife Protection Division, Department of Public Safety. Our duty was to reduce illegal harvest and waste and safeguard fish and wildlife habitats.

Biologists in ADFG acted as game wardens beginning with statehood in 1959 until 1972 when the Division of Fish and Wildlife Protection was created.[21] I was told by older colleagues that ADFG biologists retained their 20-year peace officer retirement after 1972 because so many of us die in the line of duty. By 1981, the year I was deputized, twenty biologists had been killed on the job, primarily from aircraft crashes during low-level surveys.[22] ADFG biologists had a higher rate of death in service than the troopers. There were several instances during my career with ADFG when I feared I would join the somber ranks of those who had died while conducting fieldwork. Regrettably, politicians decided biologists hired after 1984 would fall into a 30-year retirement system, reasoning there was no further need for the deputization of biologists. Although the responsibilities of 30-year retirement biologists did not include enforcement, their risks on the job were the same as always.

After disembarking from our flight to Sitka, we were transported to the academy on Sawmill Creek Road and settled into a dormitory. As we sat in the lecture room on our first morning at the academy, the corporal pointed out how seriously we should take enforcement by relating a tale of criminal behavior and his victim's fate. The details were gory. He made his point and got our attention— enforcement was serious business.

Each morning we listened to lectures on such topics as fish and game regulations, how to issue and file a citation, and preserve evidence. In the afternoon, we moved to a gym where we practiced defensive tactics and participated in simulations. In one defense simulation, the instructor grabbed me from behind, encircling his arms tightly around my chest. I quickly maneuvered out of his grip by stomping his foot, jamming his elbow upward, ducking under his arm, and twisting it as I pinned it behind him. It felt great to discover I could protect myself in this way! Sometimes we practiced on the firearms range. After passing my exam, I was assigned Badge #400. I was required to patrol periodically with a Fish and Wildlife Protection trooper to keep my enforcement skills sharp.

That fall, I wrote a proposal for a radio telemetry study of salmon in the Copper River. Ken loved the idea of using radio tags to monitor salmon movements, identify spawning areas, and estimate proportional abundance. I lobbied for funding in 1981 and 1982, but leaders had other priorities. It would not be until eighteen years later (1999) when biologists under my supervision in the Sport Fish Division initiated radio telemetry studies of salmon in the

Copper River.[23] Occasionally, ideas must wait for time to catch up with them.

While my radio telemetry proposal to study salmon was not funded, U.S. Fish and Wildlife Service biologists had funds to track steelhead trout with radio tags. Their objectives were to identify overwintering and spawning areas of steelhead in the upper Copper River, the northernmost stocks in North America. The project leader, Carl, asked me if I'd like to help them. I said, "Sure." In September, Carl, along with Butch from the Glennallen office's Sport Fish Division, and I, captured several steelhead migrating up the Copper River using fish wheels anchored near Copperville. I have small fingers and was faster at surgically implanting the radio transmitter and suturing up the incision, so the guys left that chore to me.

I cut a 1-inch-long line in the skin, slipped in the tubular-shaped radio transmitter with the antenna trailing behind the fish, sprinkled antibiotic powder into the wound, and sewed up the fish. Carl asked if I would do some aerial tracking flights. I said, "Sure." Between the fall of 1982 and the spring of 1983, I carved out time to climb into an aircraft, adjust my headphones and tune the receiver to the frequencies emitted by the transmitters. The pilot flew low, combing the Copper River and its upper tributaries, as I sought a radio signal. It was exciting to finally hear a faint "beep beep" sound because I knew we were near a radio-tagged fish. I told the pilot to alter his course to pick up as strong a signal as possible, thus pinpointing the fish's location. My flights led us to discover spawning locations of steelhead trout within the Tazlina and Gulkana drainages. Populations on their northern edge were thought to be sparse and unproductive, so additional knowledge about their natural history would help conserve the fish.

The second summer at Miles Lake, we arrived in Cordova on May 10. At mile 43 of the Copper River Highway, it was no surprise to encounter snow drifts that blocked the road. The remaining 7 miles to camp were traversed using an all-terrain vehicle. By May 19, there was still so much ice in the river we couldn't effectively beam the sonar to count fish. The fishery opened on May 17, but most of the gillnet fleet stayed at the dock because they hadn't negotiated an acceptable price with processors. I attended a fisherman's meeting in town; 25 percent of the fleet was there. I gave a presentation on our sonar project and other proposed salmon research. It did not escape my notice that I was the only woman in a room of 100+ men.

Two days later, the fleet arrived at a satisfactory price with processors and began to fish under terrible ocean conditions. High waves capsized heavily laden boats, and unfortunately, there were two drownings. Sonar counts finally started on May 23, but heavy ice conditions limited counts to only a few hours. On May 24, we were fully operational. Counts were initially low but gradually increased over the next week, indicating the run was just beginning and we had not missed too many fish. On May 27, a slug of fish arrived at the sonar counter, averaging 2,400 fish an hour. The daily count exceeded 12,000 fish. It was fast and furious. Catches were progressing at an unprecedented rate, and the sonar counter was burning up with large numbers of fish passing by.

Between count shifts, we worked on repairs and improvements to the sonar setup on both banks. On May 29, 1982, my journal read, "Camp is running more smoothly. Still have to carry supplies in on a track rig, but we're hoping the snow will melt soon so we can get a truck through. I helped pour thirty-four bags of cement and the gravel that goes with it to repair the permanent substrate on the south bank the icebergs damaged last year. Today it is sunny, songbirds are singing, and the trees are thinking about turning green."

The north bank electronics needed to be housed in a sturdier structure, so we purchased building supplies in town, trucked them to camp, and ferried them by skiff across the Copper River. One day, I loaded the skiff with building supplies and motored across the river toward the north bank with a crew member in the bow. Suddenly the skiff jerked with a thud—the propeller had hit a submerged rock, hidden beneath the silty water, and had sheared off the cotter pin. The cotter pin holds the propeller nut in place and keeps the propeller from falling off the shaft. We lost the cotter pin, the nut fell in the river, and the propeller was hanging by a thread, contorted. The current took us about 7 miles an hour toward the 300-foot-high face of Childs Glacier, which was just 250 yards away. We had less than a minute before the river carried us into the glacier. A second of panic! I was going to be smashed into an actively calving glacier and crushed to death by falling ice!

No, I was going to throw the anchor out. The anchor scraped along the river bottom. I was praying for it to find purchase when the line became taut. The anchor had wedged against a rock; please, God, it would hold. There was always a toolbox in the skiff with spare parts, something I learned from my boating adventures in Kotzebue Sound. I found a spare cotter pin and nut in the toolbox and a pair of pliers, placed them in my coat pocket, and raised the shaft out of the water. I leaned over the stern, my upper body hanging over the river, as I delicately affixed the replacement nut and cotter pin without dropping them in the river. I told myself that my fingers must work, they must not fumble, they had to hold onto the parts, and I mustn't lose my balance and topple into the river. The operation was a success! I lowered the shaft back in the water, started the engine, pulled the anchor, and sputtered back toward the south bank with a bit of uncertainty, given the damaged propeller. Again, I figured I had cheated death.

Ken designed pilings to prevent the 8 x 8-foot shed we were building on the north bank from washing downriver. The pilings were made from rails from the old Copper and Northwestern Railway, cut with a torch, set in concrete, and bolted on 4 x 12s to support the floor. My June 25 journal read, "Hot and sunny at camp, very unusual. I've kept busy working at the camp with the crew. I've learned how to frame a wall and build roof trusses."

One day when a technician and I were working on finishing the inside of the north bank sonar shed, we heard a loud crack. I knew what it was and instantly hit the floor as I told my crew member to do likewise. Someone was shooting at the shed with a high-caliber rifle from the Million Dollar Bridge! It was not unusual for locals to use ADFG structures for target practice, but I was surprised they would do so with people inside. Did they not know we were there? Surely, they could see our skiff pulled up on the bank. Were they *trying* to hit us? I certainly wasn't going to raise my head and look out the window. Five rounds hit the shed, then silence. We lay on the dusty wooden floor for a long time, waiting. I don't know how much time passed, but I finally slid over to the door, opened it, and crawled out, keeping low. I peeked around the side of the shed and looked up at the bridge—no one was there. We left the work on the north bank for another time, pulled the skiff to the water, and motored back to camp. I admit I was a bit jittery the rest of that day. I never did find out who the shooter was, but thankfully, it never happened again.

During the second season at Miles Lake, I turned thirty. I had taken the truck into town on my birthday, and one of my chores was to bring back a 50-gallon fuel oil drum for the stove. It was hammering rain, of course. I was by myself, and to get the drum into the back of the pickup truck, I set two boards on the tailgate like a ramp and rolled the drum up into the bed. By the time I was done I was muddy, soaking wet, and disgusted. This is not what I had envisioned my life would be like at age thirty! Still, I had a steady job, good friends, and good health. One really shouldn't ask for more. When I rolled back into camp and opened the cabin door, the crew had a surprise birthday party waiting for me. They even baked a cake and sang a birthday song. Kris, a fishery technician, gave me a little sewing kit that I treasured for years, and Charlie gave me a book. Wow. My crew made it all worthwhile. Thanks, guys—you salvaged my day!

Sonar counts ended on August 5. The total count was over 467,000 fish and exceeded the desired upriver escapement goal by 100,000. The commercial harvest of 1.2 million sockeye salmon was the highest since 1919 and the third highest on record. The district's catch of 49,000 Chinook salmon set a record. The estimated value of the 1982 sockeye salmon harvest in the Copper River District (based on the price per pound paid by local processors to fishers) was $7.88 million, and the Chinook salmon harvest was close to $1.85 million.[18] For a second year, I was lucky—there were lots of fish and everybody was happy because they made lots of money.

During my third summer at Miles Lake, sonar was operational on May 23 after being delayed by river ice conditions. Daily counts rapidly increased from 3,310 to 11,587. On May 31, we saw 17,123 salmon pass by. In the first week of June, cumulative escapement was at or above desired levels for that stage in the migration.

On May 29, my journal read, "There are vast amounts of snow at camp, and there's a 20-foot drift by the bridge. I'll be walking on snow until late June. We went way over budget to get the road plowed to the river but didn't have enough money to plow to camp, so we hauled supplies by hand over a quarter of a mile in deep snow. We have to try to save money, so I only drive to town once a week. We have a little shower stall out back where we pour hot water into a 5-gallon can with holes in the bottom. Despite the snow and lack of anything green, there are lots of songbirds around."

I had a good crew, and we all got along, but I had to fire one of the technicians because he was drinking on the job. I gave him one chance to stay sober. The second time I caught him drunk, he had to go immediately. I had him pack up his gear, and I drove him to the bunkhouse in town that evening, so he could begin to make arrangements for his transport home. Alcoholism is a sad condition, and while I had empathy for the man, the accuracy of our sonar counts must not be compromised or questioned. The credibility of fishery managers with the fleet and the livelihoods and subsistence of hundreds of fishers, depended on a solid sonar project.

In my three years at Miles Lake, I learned a lot about being a supervisor of a crew living in the wilderness, subject to risk and boredom. Clear and frequent communication was vital so people knew what was expected of them. At the same time, people appreciated being given a break, as we all make mistakes. Keeping morale up in a camp of diverse personalities during prolonged bouts of crummy weather was challenging! Diversions helped. For example, one year, a fishery technician, Janelle, brought a loom with her, and she taught us how to weave something during our time off. It was a fun activity to share. I still have a placemat I wove at the Miles Lake sonar camp.

I'd say that I learned two key things about supervising a camp. The first was to focus on the mission; brooding about another's ill behavior then becomes insignificant and seems a waste of time. Second, I could only inspire people to work hard in bad weather and under difficult circumstances if I'm with them every step of the way, stressing the importance of our shared accomplishments. A Chinese proverb says, "To lead people, walk beside them."

Sonar counts ended on August 4. The total count was 545,724 fish, 17 percent greater than the previous year and 56 percent greater than the desired escapement of 350,000.[24] The 1983 commercial harvest in the Copper River District was 615,000 sockeye salmon, which was above the 10-year average. The district's catch of 50,022 Chinook salmon set a new record. Prices paid in 1983 were a little low, which diminished the value of the above-average harvests. The estimated value of the 1983 sockeye salmon harvest in the Copper River District was $3.56 million, and the Chinook salmon harvest was $1.4 million.[25] For the third year in a row, I was lucky—there were lots of salmon in the Copper River, and new records were set.

Life in the Copper River Valley

Hunting, fishing, and trapping are core cultural values of rural Alaska, and the Copper River Valley embodies this spirit. More than just a means to obtain a healthy source of food, living off the land requires dedicated time, a particular skill set, and knowledge of seasonal patterns in animals and weather. Building on experience gained from previous years in the bush, I continued to hone my hunting, fishing, and trapping skills in the Copper River Valley, as much for professional as personal reasons. By spending time in the field to discover where local animals lived, how they traveled and when, what they ate, and their behavior, I would enhance my biology education and have an opportunity to add a grouse or rabbit to the crock pot. I had an additional incentive: by participating in these outdoor activities, I would have a greater chance of gaining acceptance and credibility by the predominately male cadre of decision-makers in local advisory committees.

Some of the guys in our office trapped, and after strongly hinting I'd like to learn more about it, the area biologist, Bob, finally agreed to a partnership. If I checked and maintained his 12-mile trap line near town, it would free up his time to establish a second trapline farther north toward Paxson and focus on marten. In exchange for letting me run his line, I would give him 50 percent of any fur animals I caught. After showing me the line and how to set traps, I was often on my own. I sold a few hides to Hudson's Bay Company in Ontario, Canada, but kept most of my share and used some for a coat.

When the temperature was -20° F or above, I checked the line twice a week— during daylight on the weekend and then mid-week, after work in the dark. My headlamp and flashlight gave enough light to see by, and sometimes the moon was bright. I'd pull off onto a small road, unload my snow machine from my trailer, hook up the sled and set off. At the fox sets in an open field, I'd check for tracks, sprinkle fresh urine lure on the stick, and motor on to a wooded glen where a few mink traps were located. We also set up some lynx traps. Farther down the trail, we had once spotted wolverine tracks and had made a cubby to entice the animal. In the three winters I ran that trap line, the clever wolverine always managed to snatch my bait and escape unscathed.

I felt confident venturing into the wilderness on a snow machine by myself. An entry in my journal dated January 7, 1983, read, "It's -5° F and snowing, the trap line is slow, but I love getting out. Got a mink last Sunday." A couple of situations arose that were annoying. For example, one night, as I puttered along the trail, my headlamp revealed six moose standing before me, blocking the trail. I didn't want to get off the hard-packed path and flounder in deep, soft snow, so I sat there a while, thinking. The moose just stood there, staring at me. Finally, I decided my only option was to go forward, so I puttered slowly toward the moose. They didn't act aggressive or frightened, nor did they move. They looked like they owned the trail, which is typical of moose. I swear I passed within a hand's breadth of a moose as I pressed onward, holding my breath. Made it! Thankfully, on my return, the moose were no longer there.

Weekends often found me tromping through the woods, exploring the valley. Spruce grouse were spotted in the mixed spruce and birch forest, while willow ptarmigan were discovered in the snowy hills above Paxson. An entry in my journal, dated April 20, 1983, read, "We tried to go ptarmigan hunting, but 40 mph winds, blowing snow, and hail spoiled the hunt." I bought a Mad River Kevlar canoe for duck hunting in nearby lakes; the canoe handled great and proved a stable platform from which to shoot.

One fall, I shot a caribou on the northern edge of the Wrangell-St. Elias National Park. Residents of Glennallen were deemed eligible for subsistence hunting by the National Park Service. I parked my car off Nabesna Road, hiked into the park several miles, then sat and studied the terrain using binoculars. It was a beautiful sunny day. In the late afternoon, I spotted a few stray caribou and focused on a small one with antlers; I had to be sure it was a bull because females also sport small antlers. Once I was certain, he went down quickly with one shot from my old Remington .30-06, and I commenced the skinning and quartering process. Daylight was waning, and I realized I would get only one load back to the car before night fell. I didn't relish packing the second load of meat at night over tundra tussocks with bears around. Thankfully, one of the park rangers I knew had seen my car while he was on patrol and walked into the park to see what I was up to. He arrived just in time to help me carry the meat to the car. Of course, he received a large roast for his trouble.

During my time in Glennallen, I successfully completed two hunts judged to be among the most difficult in North America—Dall sheep and mountain goat. I embarked on these hunts for two reasons. One was the personal challenge of testing my physical and mental stamina and outdoor skills that these hunts demanded. A second reason was to prove a point. In my era, I worked in a job that men had traditionally held. Some male colleagues questioned a woman's capability to do the job. A successful outcome from these hunts might help to alleviate concerns about myself and, hopefully, other women biologists who would eventually follow.

My Dall sheep hunt in the Wrangell Mountains was tough. I had prepared for it by running 2 to 4 miles daily and doing pull-ups and push-ups. While the exercise was undoubtedly helpful, in the end, it came down to a matter of grit. At the end of August 1981, a friend and I flew with an air taxi service from a lake off the Nabesna Road and landed near Jacksina Glacier, where we made camp. The next day we climbed up the base of Tanada Peak, scrambling over large boulders and fording icy streams. My friend got his sheep on the first day. It was another five days before I got mine. I focused my efforts on peaks north of camp that rose to a 7,000-foot elevation. These ancient volcanoes had eroded into jagged, crumbling cliffs over millions of years. From the base, we spotted several white dots against the black rock about three-quarters of the way up. The only way to surprise a Dall sheep is to position yourself from above. You can't start climbing below a sheep thinking he'll stand and wait for you.

We crossed the valley floor at dawn and scouted a way up the mountain. With my rifle slung across my shoulder and a pack frame hitched around my waist, I clawed and scrambled to the summit.

The climb took more than half the day. A full curl ram came suddenly into view just below, spotted me, and bounded off.

I took my shots standing and was none too steady, but after a volley of bullets, he lay dead. There was a sheer cliff between the animal and me. I sat quietly on the cliff's edge, legs dangling, trying to summon up the courage to flip around and lower myself down. The toe holds were just an inch of rock here and there. My fingers bled from digging into the rock, and my arms ached, but I somehow managed to free-climb that sheer face without falling. My friend lowered our rifles and pack frames using a rope, then clambered his way down the cliff. By the time we reached the sheep, carved it up, and divided the load, we were frantic to hasten off the mountain before it was pitch black. In the twilight, we made it to the valley floor and built a fire next to a small tree—its few branches were our only shelter for the night. I hugged my knees to stay warm and waited for light. At dawn, we headed across the valley floor toward camp. My feet were cold and wet; I couldn't feel them. It was so good to see my tent!

The pilot arrived the next morning on schedule and loaded the meat, then told me to get in—that constituted a load. He left my friend with his tent, sleeping bag, stove, and food. Because sudden weather changes may prevent pilots from returning as planned, those remaining behind are left with survival gear. On our flight out, the wind picked up. After dropping me off and unloading the meat, the pilot flew back to camp. As the plane neared the pass, a wind shear flipped it upside down, so the floats were on top! Miraculously, the pilot had enough elevation, room to maneuver, and the presence of mind to right the plane and head back. After he landed, he told me that, depending on the weather and his time, it would be a few days before he could pick up my friend. We weren't the only hunters he had to retrieve from the field. I drove to Devil's

Mountain Lodge, booked a room, put the meat on ice, and waited. I felt guilty sleeping in a soft bed and eating restaurant food while my friend was stuck in the mountains. Two days later, the pilot picked him up. Meat so dearly earned always tastes good.

The mountain goat hunt occurred at the end of August 1983. A friend who lived in Anchorage accompanied me to Valdez, and we met a floatplane pilot, Ken Sumey, who owned Columbia Air Taxi. Our destination was a snow-capped peak within the Chugach National Forest. As we shifted gear into the plane, I noticed that the pilot had water pumps on the floats because they were taking on water. I was uneasy about the leaky floats and voiced my concern. He said he had been so busy that he hadn't gotten around to repairing the cracked floats, but we would be fine for our short trip.

It was sunny when we landed on Silver Lake, milky jade in color, surrounded by coastal rainforest. We set up camp near the lake and the next morning turned toward a dreaded bush-whacking climb through noxious Devil's Club, which grows over five feet high, has leaves a foot-wide, and sharp spines. The bottom half of the steep mountain was covered by dense stands of this large understory shrub; the summit was composed of rock scree covered with patches of snow.

The medicinal uses of Devil's Club are renowned among native peoples who prepare the inner bark as a poultice, ointment, or tea for such maladies as skin infections, burns, and pneumonia. While I appreciated the plant's lofty reputation in traditional medicine, I did not appreciate its ability to rebuff my attempts to advance through it. My boots slipped on the tangling stems. The spines snagged my clothes, caught my pack, and scratched my skin as we struggled toward the summit. At last, breaking out into open rock, we continued upward until we found a good vantage point from which to glass the mountainside and wait. We hid motionless behind a blind of rocks for hours. Then, I heard a faint ripple as grains of dirt tumbled beneath a goat's hooves. A small group came into view. We both had sure shots and felled two large goats instantly. The tricky part was getting to them. John's goat was fairly close by, but mine had tumbled down a steep chute of loose rock that could avalanche.

We field-dressed John's goat, loaded his pack, and slid down the slope to my goat. After loading my pack and standing up, I discovered that mountain goats weigh more than Dall sheep. Hmm, my pack weighed at least 150 pounds. Thank goodness the return to camp was all downhill. I slid down the rock chute on my heels and butt, then entered the thick understory brush.

Me with mountain goat

The lower mountain was steep and slippery. With the weight of the heavy pack pressing on my feet, my toenails turned black and later fell off. By the time we reached the forest floor, it was dark. Then, a terrible pain shot through my side, and surprised, I collapsed to the ground. I suspected that my abdominal muscles had ripped and created a hernia. Holding my side, I continued walking to camp. We tied several bags of meat and the capes high in a tree with rope to keep the black bears from getting into them. I was grateful to sag into an exhausted sleep.

Shortly after midnight, we heard a black bear in camp. The bear had scaled the tree with the meat and scampered away with a bag when we fired a warning shot. Sleep was fitful after that as we took turns watching for the bear's return. It rained the next three days. At the appointed hour, the pilot returned, and we departed Silver Lake.

Two weeks later, back in Glennallen, I received a distraught phone call from John. The floatplane pilot in Valdez had crashed and killed himself and his hunter! The September 28, 1983 issue of the *Valdez Vanguard* read:

> A Valdez pilot (age 34) and a hunter from Anchorage (age 22) are presumed drowned after a floatplane they were in last Wednesday, September 21, apparently sunk in Silver Lake. According to Alaska State Trooper Jim Alexander, goat hunters in the area reported that at about 11:15 a.m., the plane was circling the lake, getting ready to land, when the left float hit the water rather hard. One of the floats broke off, and the plane was thrown up in the air and then dropped into the middle of the 300-foot-deep lake. Witnesses said there was no wind, and the surface of the lake was smooth. The two men were able to get out of the airplane underwater, Alexander said, adding that the temperature of the water was 38°F.[26]

The article went on to say the men clung to the tail of the plane. One man tried to swim to the disconnected float, then gave up and swam back to the tail. The men hung on for two hours until the hunters on shore couldn't see them anymore due to heavy rain. The hunters on shore frantically tried to build a raft out of trees to try to reach the plane, but they were unsuccessful. The lake's silt level thwarted efforts to recover the bodies. Both men left behind families with small children. I felt awful for the victims and their families. The fact that we had been in that same plane, at that same lake, just two weeks earlier caught my breath. That could easily have been us! Either fate or a guardian angel had intervened to spare my life again.

***

During my first week in the Glennallen office, I was invited to train and serve as a volunteer with the Copper River Emergency Medical Service (CREMS). My boss, Ken, was a volunteer firefighter and mentioned more volunteers were needed in EMS. I said yes; volunteering would be a great way to make a meaningful difference in people's lives in the Copper River area. My availability to volunteer was restricted to spring and winter due to my commercial fisheries work in summer and fall. Faith Hospital, located in Glennallen, served as the only healthcare facility for people living in the fifteen tiny communities nestled within the Copper River Valley.

The service area stretched from Nelchina in the west, Paxson in the north, Mentasta Lake in the east, and McCarthy in the south—roughly 10,800 square miles in size, as big as the state of Massachusetts. In addition to the vast geography of the service area and the remoteness of the communities, winter weather conditions posed an additional challenge to emergency responders.

One of the two doctors who staffed the small hospital, Ross, was offering an eighty-hour Emergency Medical Technician (EMT) Level I course through the Alaska Bible College, affiliated with and adjacent to Faith Hospital. I joined the handful of locals who signed up. Classes met at night, six hours a week. Trainers from outside the community were sometimes brought in, such as paramedics with the Anchorage Fire Department. They demonstrated the safe extraction of victims in car crashes, including the operation of the "Jaws of Life," a hydraulic tool that cuts and spreads metal.

During the course, we were encouraged to ride in an ambulance as observers to better understand what to expect and how our training would be used. When I was called into the Anchorage office for commercial fisheries business, I rode in an ambulance with the Anchorage Fire Department in the evenings. Their calls were more numerous and often of a different nature than the emergencies we responded to in the Copper River Valley. Working during the day and studying at night helped make the long Alaskan winter go by. Upon passing our written and practical exams, we were certified as EMT Is and issued pagers that we wore on our belts.

As 911 calls came to the troopers in Glennallen, the calls were relayed to the hospital staff, who notified volunteers on call for that shift. I was usually on call forty-eight hours a week. Volunteers on call were supposed to stay within a 10-minute driving radius of the hospital. When our pagers beeped, we drove to the hospital, were briefed on the emergency, given directions, and jumped into the ambulance. It was not unusual for the ambulance to take an hour to reach victims located in the far-flung edges of the service area. Per protocol, I kept a log of all my calls, both with CREMS and the Anchorage Paramedics, whom I continued to ride with during the three winters I was an EMT in the Copper River Valley. In the log, I noted the date, responder names, type of accident, description of the victim, symptoms, treatment, and response. I sometimes noted follow-up information about the victim's condition from hospital staff.

My first ride in an ambulance was on December 16, 1981, with the Anchorage Paramedics. The call was a motor vehicle accident involving a woman, age 80, with minor head lacerations. She was covered with hundreds of tiny glass fragments, which I gently brushed off her clothes and face. Her pulse and blood pressure were normal, and she refused treatment, insisting she was okay. The next vehicle accident involved a drunk driver, and after transporting him to the hospital, I noticed a police guard was assigned to stay with him in the emergency room. Another vehicle accident occurred, again with a suspected drunk driver, who was taken directly into police custody. I learned that drunk drivers are a frequent cause of emergency calls. Three years later, a drunk driver would forever change my close-knit EMS community and influence the rest of my life.

In Anchorage, a week later, I rode with the Anchorage Paramedics again, and we had a busy night with six calls. One was from a poor woman who fell out of her wheelchair. We had to break the door down to gain access to her. Several more house calls followed, which can be dangerous if determined to be domestic violence. In that case, the paramedics request police backup. The paramedics coached me to be defensive, aware of who was behind me in a house and where the nearest exit was.

I am haunted, to this day, by a call we made to an Anchorage nightclub. A young woman, twenty-three, had been slipped a hallucinogen in her drink. She was so scared. "What is happening to me?" she cried. Her nervous system was fried. Very quietly, the paramedics moved her into the ambulance with lights turned low;

they talked in a low monotone and ran quiet (no siren). Any stimulation sent her into a screaming hysteria. They had to apply restraints. The paramedics were her only tether to reality as they gently held her hand and whispered reassurances as she writhed and screamed in fear. Doctors ran a drug test in the emergency room and confirmed what the paramedics had suspected: PCP or "angel dust." How could anyone be so cruel as to send another person into a living Hell—an innocent young woman without her knowledge or consent. I vividly met evil that night. Years later, with that emergency call in mind, I cautioned my daughter about being careful when going out for drinks.

The first ambulance ride I made as an EMT with the CREMS was on January 11, 1982. We were called to the scene of a home propane bottle explosion. A man, age 75, had sustained second and third-degree burns on his hands and face. We applied saline-soaked dressings and transported him to the hospital. Next, an elderly man was picked up from his home in shock. I could not detect a radial pulse, and his blood pressure was low. He was medevacked to Anchorage, where he was diagnosed with kidney failure.

There were victims of lacerations incurred during drunken brawls, migraine, miscarriages, strokes, heart attacks and motor vehicle accidents during my first winter. Treating victims of motor vehicle accidents in winter was difficult. While we needed to extricate and stabilize with care, the freezing temperatures made everyone miserable and confounded symptoms. Was the patient trembling from cold, fear, pain, or an underlying neurological disorder? On September 23, 1982, my first patient died. He was a cancer patient, and after transporting him to Faith Hospital, we all worked on him in the emergency room for two hours, to no avail. His death made me sad. We had tried so hard to keep him alive.

After dealing with an extrication from a car that had gone over a cliff, I felt strongly that in our remote and rugged country, volunteer responders needed training in mountain and avalanche rescue techniques—not something we would typically receive in an EMT course. Accordingly, in March 1982, I took time off work to attend a four-day hazard evaluation and rescue workshop taught by the Alaska Division of Parks. I spoke to the instructor, Doug, about the need for rescue training in the Copper River Valley, and he kindly agreed to come to Glennallen and give our CREMS squad a twelve-hour workshop in rescue techniques. Members of several government agencies in the valley also participated in the workshop. Doug taught us how to belay down a steep slope to a victim and secure and haul up a basket.

I noticed there was no coordination among local, state, and federal agencies for search and rescue in the Copper River Valley. The troopers tended to rely on search and rescue groups outside the valley, increasing response time and thus decreasing the chance for a successful outcome. To improve the situation, I worked with local and trooper coordinators to organize the first Copper River basin interagency search and rescue meeting in April 1982. In attendance were local representatives of the troopers, ADFG, CREMS, Bureau of Land Management, Alaska State Parks, and Alaska State Forestry. In subsequent meetings, we were joined by members of other agencies and private pilots. This was a great response! At the meetings, we identified the equipment, expertise, and staffing available for a mission, and best methods of communication.

On April 22, 1982, I responded to a CREMS call to transport a twenty-one-month-old baby from Copper Center who had a suspected case of meningitis. This was the second baby from Copper Center with meningitis. The first baby had died from it a week earlier. I was nervous about exposure to this disease. The symptoms of meningitis are headache, vomiting, fever, and stiff neck. After transport, we washed the interior of the ambulance and changed the linen, but I worried we hadn't taken enough precautions. No other cases of meningitis were detected that winter. The call was a sobering reminder of the diversity of risks that EMT volunteers face.

My second winter of emergency calls as an EMT with CREMS began in late August 1982 with a seizure victim. Two weeks later, we responded to two male cardiac patients. One man made it, and the other didn't. The second patient had really low blood pressure and a rapid pulse. We administered oxygen at 10 liters/minute, put on Medical Anti-Shock Trousers to increase his blood pressure, and raced to the hospital. We stayed with him as the doctor worked on him, but his blood pressure never recovered. I was standing next to him when he died. All of us tried, but the doctor let us know that sometimes, trying our best would not change the outcome—it was beyond our control. The winter progressed with calls for a baby with burned hands, a migraine, a tubal pregnancy, a miscarriage, and congestive heart failure.

On March 1, 1983, I was on call to go with the ambulance to Gulkana Airport and pick up a male, about age 30, with a gunshot wound to the head. A small plane was bringing him in from McCarthy for stabilization by doctors at Faith Hospital before medevac to Anchorage. Our ambulance arrived first, and we waited for the small plane to land and taxi to a stop. As we reached inside the plane to retrieve the patient, I was in awe of his lucidity despite the gunshot trauma. He was upright and talking fast as if talking was the only thing keeping him from slipping into unconsciousness. He had a gaping hole in his head, from which the bleeding had slowed. His blood pressure was good. Thank goodness he was not in shock. I re-bandaged his wound as we raced to Faith Hospital for evaluation by doctors. The patient was named Chris Richards, a survivor of what later came to be called the McCarthy Massacre. I held Chris' hand in the ambulance, listening, nodding, locking eye contact, keeping him focused on me, urgently willing him to continue breathing, to continue talking, to continue living.

Later, I learned the details of his attack. Chris was one of two people living in the ghost mining town of Kennicott, in the heart of the Wrangell-St. Elias National Park. He had settled there in the 1970s. In the summers, Chris guided tourists through the historic buildings. The nearby town of McCarthy was home to about two dozen people, neighbors and friends of Chris. A newcomer had entered this close-knit community named Lou Hastings. On the morning of March 1, Chris heard a knock on his door and called for Lou to enter. As Chris reached for the coffee pot, Lou pulled out a Ruger Mini-14 semi-automatic rifle and fired a bullet into Chris' cheekbone. Chris struggled with Lou for the gun and stabbed his attacker in the leg with a kitchen knife, then fled through the waist-deep snow to the cabin of Tim and Amy Nash. Chris was transported by snow machine to the McCarthy runway, where a local pilot loaded him into his plane and took off, radioing for help. Tim and Amy Nash armed themselves and went to warn others. While Chris was being flown to Glennallen for medical care, Lou hunted down and killed six people, including the Nashes. A trooper helicopter chased down Lou as he tried to escape on a snowmachine. Lou was sentenced to 634 years in prison for his crimes. Chris recovered from his wounds and returned to Kennicott.[27]

While I sometimes responded to calls by myself, most of the time, I had a partner who was another EMT or a nurse with Faith Hospital. In the winter of 1983–1984, I was accompanied on many calls by Peri McIlroy, an EMT who became a friend. Peri had two children, Kara and Kurt, who were finishing high school. On January 30, 1984, in the depths of a frigid -40° F temperature, we received a call to help a truck driver. Three EMTs responded: me, Peri, and our EMT chief, Kathy, and in an unusual move, Ross climbed into the ambulance. The doctor was tight-lipped about the nature of the call. It took thirty minutes to reach the site of the emergency. As we piled out of the ambulance and approached the parked truck, Ross held up his hand and told us to stop. He went to the truck alone. Ross knelt, got partially under the truck, then rose and came back to talk to us. Apparently, the victim had gone under the truck to make an adjustment or fix a malfunction, and the rapidly-spinning power take-off (PTO) shaft caught his coat sleeve. The PTO quickly entangled his arm around the driveshaft and ripped it off, spinning his head into the underbody, instantly killing him. We all felt terrible because we knew the victim and his family.

I was invited by Billie Peters, Health Director with the Copper River Native Association, to serve on the Copper River EMS Council Board of Directors and was elected vice president. I was kept busy with ambulance runs, search and rescue meetings, refresher EMT classes, and the CREMS Council. I devoted nearly as much time to volunteering as I gave to my day job! I realized that rewards were found in the comfort and relief that I could provide to people who were injured and scared. By giving, I felt alive. I also enjoyed the friendships formed with the CREMS crew and hospital staff. Invitations went back and forth between homes to gather for celebrations, with music and great food.

***

When I moved to Glennallen in 1981, housing options in Glennallen were limited, so I was fortunate Ken had arranged for me to rent a room in a large home owned by Joni. The house was close to the

ADFG office. Joni's decision to sell her house and leave the valley for further schooling provided the impetus for me to look for my own place.

I began looking for houses in the Glennallen area. There were few for sale, and all were junk. I considered buying land and building and found a fantastic piece of land in Gakona, 11 miles north of Glennallen. Wild raspberries and roses grew on the property that bordered a pretty creek. The parcel was on a south bluff with a view of the Wrangell Mountains, and had a road and power. I envisioned designing a beautiful A-frame cabin with an open beam ceiling on 10 acres and getting the foundation and frame up before winter. Reality set in once I began gathering contractor estimates, which were higher than expected. Another stumbling block was that I would be gone all summer at Miles Lake during the building season and unable to oversee construction.

Three days later, I heard about a three-bedroom house for sale within walking distance from the office on a wooded 2.5-acre parcel next to Moose Creek. The old couple wanted to move to Oklahoma. The house came with appliances, furniture, and outbuildings. When I walked inside to inspect the house and meet the owners, I noticed a distinct tilt—yes, a corner of the foundation rested on melting permafrost. The foundation needed to be addressed, and the well needed an upgrade before the National Bank of Alaska would give me a loan. There were no other housing options.

Before I made a significant property investment, I considered my commitment to my job. I loved the country. My boss and I got along well, and he gave me opportunities to participate in various research projects. I enjoyed the friendship of my office family. I was thrilled to have attained permanent employee status. So, that settled it! I would have my own house to putter in during the winter, and I could rent two bedrooms to help with expenses. The owners wanted $65,000, $6,500 down, and 12.125 percent interest. I turned in my loan application in October but did not sign the papers until three months later, in January 1982.

My house in Glennallen

I moved in during a cold snap of -50°F. Those cold temperatures are hard on cars. At -50°F, it took me two hours of using a propane heater under my car and a jump to the battery to get my car started, even though I was using a battery blanket, a circulating heater, and a blanket draped over the hood. The grease in the steering wheel froze, so I could hardly turn it. The tires froze flat, so it was bumpy as the tires clunked around. Thank goodness I bought a house close to town. I just bundled up in the winter clothes I'd assembled while living in Kotzebue and walked to work.

After work, I unpacked boxes and began my journey repairing and maintaining a money pit. The March 1, 1982 entry in my journal read, "I'm happy to be in my house, but this place needs so much fixing up! There is no carpet. I drove to Anchorage and got a deal on carpet at Pay n Pak, $1,300. An energy auditor with the state said I need more insulation: the ceiling will cost $1,200 to attain

R38, and I need insulated doors and thermal windows. A wood stove will cost $600. I got used furniture at Elmendorf Air Force Base: a bed for $20, dressers for $60, chairs for $15."

Soon, I had two women tenants: a National Park Service employee and a Copper River School District teacher. I charged $225 per month and half the cost of utilities. During the 1982 field season at Miles Lake, Ken kindly arranged for me to work ten days on and four days off, so I could travel from Cordova to Glennallen, at my own expense, to work on the house.

A June 25, 1982 entry in my journal read, "Flew to Valdez, got a ride to Glennallen. The house looks different with green trees. I mostly spent my time under the floor putting in fiberglass insulation, a terrible chore! I wore a face mask and goggles. The floor didn't have any insulation, and it will be worth it this winter. A black bear has been around the house, looking in the window (scared the girls). It tore up the back end of my snow machine to get at my trap lures. I like living where bears and moose roam through my yard, although I don't like the damage."

A black bear outside my kitchen window

In winter, I worked on the house or with EMS in the evenings and trapped or hunted on weekends. Paint, wallpaper, linoleum, curtains, grout, paneling, and bricks were picked up in Anchorage whenever I attended a meeting there. I hired someone to replace the sagging, stained ceiling tiles with sheetrock. My September 23, 1982 journal entry read, "Had a carpenter put in the thermal doors, but he put the back door in backward, and it will have to be redone. The toilet is worse. A valve is leaking water, I don't have time/knowledge to fix it, and there are no plumbers in town, so that's a problem. A friend and I spent three evenings trying to straighten the house by jacking up the foundation. I'm going into Anchorage next week and will pick up more stuff for the house."

I sewed Roman shades for window coverings and was especially proud of the brick base and wall I put in for the wood stove. A year later, I was still working on the house. My September 8, 1983 journal entry read, "I worked on the bathroom. The wallpaper is up, and the mirror, sink, and light are in place. Just have to connect the plumbing on the sink. I have been working hard to finish it."

Winter threw challenges at me. There was the time my propane gas for the stove liquefied, a state that occurs at -45°F. The furnace broke, which caused my pipes to freeze, then the well iced up. I installed heat lamps in the well but couldn't get any water. After working on it for days, I called the previous homeowner in Oklahoma, and he said to check a foot valve in the well—it was stuck. I got a new motor for the furnace and a pump for the well.

While labor is rewarding and hard physical work is good for the spirit, I was beginning to tire of incessant work on the house. I took small breaks for fun by joining a ladies' basketball team, sewing a quilt for my bed, and doing leather work. I even signed up for piano lessons with Becky Joy, a lady who came to the valley in 1936 as a missionary. She wouldn't take money, so I gave her parcels of game meat from my freezer. In October 1983, my house was finished, and I could relax and enjoy it!

While Alaska had prospered since oil started flowing through the Trans-Alaska Pipeline in 1977, in 1983, Governor Bill Sheffield sensed the economic boom would fade, so he cut the state budget for Fiscal Year 84. We heard rumors that four positions would be cut from the Commercial Fisheries Division in Region II. My position as a Fisheries Biologist II in Glennallen was listed as a low priority. In October, I received a call from the research supervisor in Anchorage, Chuck, who asked if I would consider another job, as openings were coming up in Soldotna and Homer. It looked like Ken would lose his assistant position, and I would need to transfer. The transfer would occur in March 1984. I had some time to consider each location and prepare for the move.

In November, Ken received a call from the shellfish manager in Homer, who said the Homer staff had heard I was being considered for the shellfish research biologist position there, made vacant by the recent retirement of Al. Al had held a Fishery Biologist III position in Homer for ten years. The manager told Ken he was concerned about a woman working in shellfish; if I accepted the position, "the staff would destroy me." Ken was shocked at this threat! Chuck admitted that Homer staff were leery of working with a woman, but when he had asked Al if a woman could handle the job, Al had said, "Yes."

With strong support from my new boss, Chuck, I decided to take the position in Homer, despite the odd warning. Homer offered an abundance of outdoor activities, and the position was permanent, while the Soldotna job was temporary. I was to be transferred as a Fishery Biologist II, responsible for performing the same duties as a Fishery Biologist III. Complicating the picture was that I would supervise an assistant, also a Fishery Biologist II. I was given verbal assurance from Chuck that I was eligible for the Fishery Biologist III job and just needed to prove myself first.

I felt vulnerable to forces beyond my control, and sad to leave the house I had *just* finished remodeling and the Copper River Valley. However, I also looked forward to new challenges, seeing new country, and learning more about fisheries research. I had built a strong core of happiness and had much to be grateful for: good health, a professional career, rewarding EMS work, and a finished house that I could lease. I figured things would work out okay. I was about to embark on an unbelievable experience in more ways than one.

Lower Cook Inlet Shellfish

"Life is not easy for any of us. But what of that? We must have perseverance and above all confidence in ourselves. We must believe we are gifted for something and that this thing must be attained." ~Marie Curie

The picturesque city of Homer rests on the shores of Kachemak Bay at the southern tip of the Kenai Peninsula, 230 miles south of Anchorage. In the mid-1980s, the town of about 2,500 people primarily relied on commercial fishing, sport fishing charters, and tourism as its economic base. The ADFG office is at the head of Homer Spit, a 4-mile-wide gravel bar that extends into Kachemak Bay and accommodates the Homer Harbor at its end. The state research vessel, the *Pandalus*, is docked here. The climate is mild, with average winter temperatures hovering near 30°F, summer temperatures in the 60s°F, and average precipitation at 24 inches.

Overlooking Homer Spit and Kachemak Bay [28]

Upon arriving in Homer, I found a vacant apartment in a four-plex on the hillside above town, with a beautiful view of Kachemak

Bay. As the Lower Cook Inlet (LCI) area shellfish research biologist from April 1984 through December 1985, my job was to assess the status of the shellfish resources and the commercial fishery's impacts on these resources. To accomplish this, a quarter of my work time was spent on the water, conducting surveys and sampling shellfish.

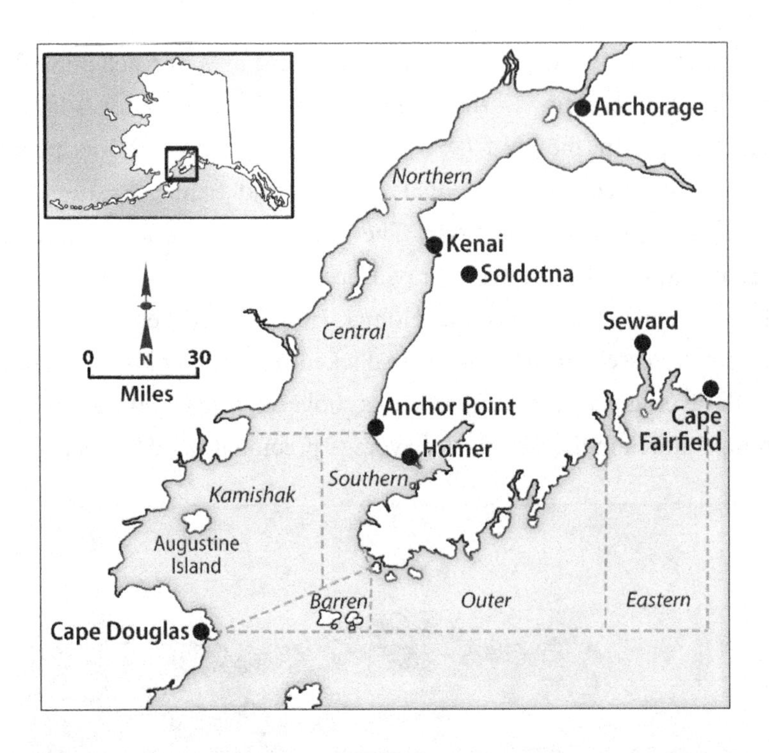

My research area was described as waters west of a line extending south from Cape Fairfield and north of a line extending east of Cape Douglas. While I spent most of my time in the Southern and Kamishak Districts, we also paid attention to activity in the Barren, Outer, and Eastern Districts. Shellfish harvested in the Central and Northern Districts, notably razor clams, were monitored by ADFG in Soldotna. In addition to residents of Homer, commercial fishermen from other Alaskan ports, such as Seldovia and Port Graham, and as far away as Seattle, participated in the LCI shellfish fisheries.

My transfer to Homer was resented and resisted by male colleagues, one of whom had applied for my position and had not been chosen. I was the first female biologist the staff had encountered, and they were leery of working with me. Three momentous changes were about to upend their routine. The first was my presence, followed a few months later by my installation of the first computer to ever sit in the Homer office. Finally, biometricians began to review and question their conclusions. Some biologists at that time considered "number-crunchers" ignorant of the "real" picture of stock status found only through fieldwork.

The office was staffed by two female secretaries who sat in the main lobby. The only single-seat bathroom for the building sat off from the main lobby. Down the hall from me was the lab where we stored, sorted, and measured samples brought back from surveys. I had a storage bin for small supplies. Large gear, like crab pots, were stored in a yard on Homer Spit near the dock. Staff with the Commercial Fisheries Division included my assistant, the area finfish manager and his assistant, the area shellfish manager, and two technicians. The *Pandalus* crew, all able seamen, consisted of Skipper, First Mate, and Cook. During surveys, First Mate and Cook helped deploy and retrieve gear. The Fisheries, Rehabilitation, and Enhancement Division had two biologists stationed in Homer, my friends Nick and Bill. No Game or Sport Fish Division staff were in the Homer office then.

Skipper announced at my first Homer staff meeting that he "didn't think a woman should be on the boat because they couldn't handle the job." In addition to their dislike of female biologists, staff had developed an antipathy toward each other: the shellfish and finfish managers were not on speaking terms; the crab technician,

Bubba, had a habit of walking away from his job if he felt like it, leaving others to pick up the slack; and, Skipper and Cook had frequent disagreements. Chuck emphasized the staff needed to work as a team and encouraged me to foster teamwork—an uphill battle!

As I became familiar with the database, I realized the shellfish manager had not written a management report since 1976. Vital data was missing or had never been collected, and files were disorganized—this made my job more difficult. Over the next year, when I had time, I keyed historical harvest and survey data into the computer to create an organized database. Not surprisingly, as I prepared for the 1984 field season, I found that the 1983 red king and Tanner crabs survey data had yet to be analyzed and reported. I worked double time to write the report.[29]

An important task of an area biologist is communicating with the public, which I did a lot of in Homer. According to my log record, 40 percent of the days I spent in the office included varying degrees of public interaction, from sealing bears to responding to questions from commercial fishermen.

I found commercial fishermen in Homer knowledgeable about shellfish resources and fully engaged in the management process. Sometimes for hours at a time, they sat in my office and discussed their ideas and proposals with me. I recall meetings with fishermen who have since passed away: Bob Prue talked about pot shrimp openings and ideas for Dungeness crab research; Bob Moss talked about red king and Tanner crabs; Chuck Parsons was interested in trawl shrimp; and Garland Blanchard was concerned about Dungeness crab molting frequency and movements. I was glad commercial fishermen in Homer felt comfortable enough to call or walk into my office, knowing I would listen and treat them respectfully. However, these discussions caused long work hours for me. I often worked thirteen-hour days.

During my first week in the office, two members of the Homer Fish and Game Advisory Committee, Jim and Dan, came to introduce themselves. After that, they either phoned or visited regularly to ask about research results, fishery openings, and give me their suggestions for research. Whenever the committee met, I was usually there to give presentations and answer questions. As a member of the committee's trawl shrimp task force, I participated in long meetings that could last until midnight. Advisory Committee members from Seldovia also stayed in touch with me regarding survey results.

One of my favorite people to drop by the office was my predecessor, Al. Although retired, he had remained in Homer. In a friendly manner, he discussed surveys and gave me his perspective. I appreciated his continued interest in the research program he had guided for many years and welcomed his visits.

Because education about ADFG's research program promotes a better understanding of the area's resources and their management, I agreed to give lectures at Homer High School and Kenai Peninsula College's Kachemak Bay campus. On a limited basis, I also agreed to allow students to come aboard the *Pandalus* during surveys in Kachemak Bay to observe sampling methods.

I was surprised at the number of requests I received from colleagues for samples, information, or collaboration. There was frequent communication between myself and the National Oceanic and Atmospheric Administration staff because they wanted me to help monitor deep-sea temperature and salinity in LCI, and I wanted access to their database. University of Alaska Museum researchers wanted mollusks and live king crabs. There were data requests from British Columbia about Dungeness crab, Alaska Sea Grant about oceanographic conditions, and Kodiak ADFG about crab abundance and fleet effort. I entered into a multi-agency collaborative effort to research crab parasites. I spent a lot of time responding to a request for information about LCI shellfish from the Habitat Division for the Alaska Habitat Management Guides. These regional maps display the distribution of species and their uses. While my professional network greatly expanded and collaborations hastened the flow of exciting research information, dealing with requests added time to my already long days.

When not at sea, I'd run on the spit during lunch hour. In winter, if Beluga Lake was sufficiently frozen, I'd ice skate, or a couple of the guys from the office and I would play a friendly game of ice hockey.

The hernia acquired on my previous fall's goat hunt was getting worse, so I checked in to the South Peninsula Hospital in Homer during the first month in my new job. I met a doctor there, Paul, who said he was a part-time surgeon, and part-time gold miner. He agreed to operate but said, "There's no anesthesiologist. Would you mind having a local administered?" I replied, "Let's try it." I didn't feel slicing or sewing, just tugging, as he gaily chatted about the Homer environs and his gold prospecting adventures. Thirty minutes after the operation, he said I could drive home and take it easy for a few days. Taking it easy was short-lived, as I had the pot shrimp survey to do, followed shortly thereafter by the trawl shrimp survey and dockside sampling of commercial pot shrimp.

Pot Shrimp

Kachemak Bay is a highly productive habitat for the five commercial shrimp species in the *Pandalus* genus: spot, coonstripe, sidestripe, northern (pink), and humpy. They differ in size, with spot shrimp attaining up to 9 inches in length while the humpy is cocktail-sized. Gyres within the bay mix land runoff with Gulf of

Alaska waters to produce a nutrient-rich base for plankton, which serves as a food source for shrimp.

The commercial pot shrimp fishery in the Southern District, primarily in Kachemak Bay, began in the late 1950s with a few vessels harvesting mostly coonstripe shrimp and smaller amounts of spot shrimp from protected bays. Harvest was low, and the shrimp were sold locally. In the 1970s, market demand increased. When 676,978 pounds of pot shrimp were taken in 1974, guideline harvest levels were established to spread the harvest throughout the year. From June through September, a total of 100,000 pounds could be taken, and from October through May, up to 500,000 pounds could be taken.

The problem was that coonstripe shrimp were also taken in the commercial trawl shrimp fishery, and adding up their harvest from both fisheries meant that a lot was gleaned from Kachemak Bay.[30] At its height, the ex-vessel value of the commercial pot shrimp fishery was an estimated $370,700 in 1977 for Cook Inlet waters. The shrimp couldn't sustain the high exploitation. By the mid-1980s, when I came on the scene, harvest levels had been slashed to 25,000 pounds in each of three seasons.

Pot shrimp index surveys began in 1978 in the Southern District during March, May, and October. I continued the index program as it had been previously implemented. In March 1984, before I was to officially begin my job on April 1, Chuck suggested I join an upcoming pot shrimp index survey to become acquainted with Homer staff and crew of the *Pandalus*.

The sea was calm. The bright sun reflected off the water and gave me a sunburned nose. The next time, I remembered to bring sunscreen. I loved being on the water, learning new sampling methods, and anticipating what each pot would bring when brought to the surface.

The *Pandalus* with Augustine Island to the right

Skipper told me the rules of the boat: no whistling on board as it would "whistle up a storm;" and no women in the wheelhouse. While I respected the first rule, the second was untenable. As Project Leader, I needed to communicate with Skipper, although I kept my visits short as I knew it irritated him. Beginning with that first survey, Cook began to harass me by asking for dates or making crude and sexist remarks.

We surveyed forty-four stations in Kachemak Bay systematically selected from a 1 nautical mile^2 grid pattern of the bay. We deployed three pots per station about 300 yards apart, for a total of 132 pot pulls. The pots were 48 x 48 x 20 inches in size and were covered with small mesh. Shrimp were enticed, by two perforated jars of chopped herring, to enter tunnels leading into the pot. Each pot had a line connected to a buoy. The pots soaked for twenty-four hours before we picked them up. The catch in each pot was sorted as to species, and coonstripe and spot shrimp catches were weighed. Samples were taken to the lab for further study.

In the evenings after coming ashore, I worked up the day's survey data, conferred with the shellfish manager, fielded questions and advice from fishermen, and measured shrimp. My assistant and I spent hours at a time in the lab; we measured thousands of shrimp carapace lengths for age and growth estimates using hand-held calipers, a tedious and time-consuming task. After the completion of each survey, a summary of the results was immediately made available to the public. Another task was to trudge down to the storage yard on the spit and mend holes in each pot's mesh before the next survey.

The pot shrimp survey results varied seasonally and were as follows. In 1984, the March, May, and September mean catch-per-pot was 8.4, 2.7, and 4.1 pounds, respectively.[31] The 1985 survey numbers were slightly lower.[32, 33] The shellfish manager thought the 1984 and 1985 survey catches were sufficient to open fishing seasons. During 1984-1985, twenty-two vessels harvested 76,100 pounds of pot shrimp, which fishermen marketed locally for $1 a pound. During 1985-1986, twenty-five vessels harvested 72,100 pounds of pot shrimp. Due to persistent low abundance, the commercial pot shrimp fishery in the Southern District was closed permanently in 1988.

Trawl Shrimp

Commercial trawl harvest of shrimp began in the 1950s in the Southern District, primarily Kachemak Bay. Most of the harvest consisted of two species, pink and humpy shrimp. Smaller amounts of sidestripe, coonstripe, and spot shrimp were also harvested.[34] By the late 1960s, annual trawl catches reached 5 million pounds and remained near this level for a decade. In the 1970s, guideline harvest levels were initiated for two seasons: 3 million pounds could be taken from mid-April to mid-October, and 2 million pounds could be taken from mid-October to mid-April.

In 1979, a Kachemak Bay Trawl Shrimp Management Plan was enacted by the Board of Fisheries. The plan sought to distribute fleet effort across three seasons; in each season, weekly allowable harvest amounts were set.[35] However, the shrimp population could not sustain the high exploitation, and by the mid-1980s, the commercial harvest had decreased to between 1 and 1.5 million pounds.

ADFG initiated research on trawl shrimp in Kachemak Bay in 1971. The goal was to monitor stock status and establish harvest guidelines with index surveys and catch-per-unit-of-effort data from the commercial fishery. My job was to continue the Southern District trawl surveys, which were done twice a year in May and October. The surveys resulted in estimates of abundance using an area-swept method of expansion, species compositions, and sex and length compositions.

My first trawl survey was May 23–29, 1984. Before the survey, I touched base with Skipper about survey design, logistics, and the roles of crew and staff. During the survey, a bracing wind turned our noses and cheeks red. I wore a wool hat, insulated gloves, a float coat over my Kelly Hanson rain pants, and Juneau boots. The air smelled of diesel mixed with fish. My legs easily adjusted to the slight swaying of the deck, unconsciously balancing in rhythm with the ocean swells.

We surveyed thirty-four stations in Kachemak Bay. Stations were selected from a 1 nautical mile^2 grid pattern of the bay that were greater than 20 fathoms (120 feet) in depth, in approved fishing locations, and away from known areas of bottom snags, underwater cables, and mud. For each station, a 1 nautical mile tow was made at a speed of 2 knots that took approximately thirty minutes. This bottom speed matched the speed of commercial shrimp trawlers. Such variables as swim speed and size of the target species and current direction and velocity impact catch. Skipper maintained consistent tow speeds to reduce sampling variability in the estimates.

A full trawl net is brought onboard [28]

The shrimp otter trawl net we used had a 61- by 32-foot opening with the mouth spread open by "doors" set at an angle on either side. As the trawl was towed through the water, it engulfed marine animals, and they tumbled into the tail or "cod end" of the net. The top of the net was held up by plastic floats connected to a "headrope." The bottom of the net was held close to the sea floor with weights tied to a "footrope." Attached to the footrope was a metal "tickler chain" that was bumped along the seabed in front of the net to disturb marine animals lingering on the bottom, causing them to enter the net. The net was deployed from the vessel's stern, attached to thick wire cables called "bridles" that were uncoiled from a drum.

After each tow, the net was hauled aboard and weighed with an electronic hanging scale before the cod end was untied and contents were dumped onto the rear deck. Total net contents ranged from a few pounds to over 1,000 pounds, with the average weighing a couple hundred pounds. Non-shrimp species were noted and returned to the sea. Our tows were punctuated with the raucous high-pitched cries of seagulls as they jostled each other overhead, eager to scoop up marine life tidbits escaping from the net or tossed overboard. The discards provided food for scavengers.

I'm collecting a random sample of the catch [28]

Two random bucket samples of 4 gallons each were collected from tows weighing 500 pounds or more; for catches less than 500 pounds, one bucket sample was taken. The samples were processed quickly as follows:

- Fish, shrimp, and debris were separated. Each group was weighed to obtain an estimated percentage of the total catch.
- For shrimp, a 5-pound sample was randomly selected and separated by species.
- Each species group was weighed to obtain an estimate of the composition of the catch.

- Because pink shrimp comprised the highest percentage of the catch, a sub-subsample was frozen for analysis in the lab.
- In the lab, the frozen pink shrimp were thawed, segregated by sex, weighed, and their carapace lengths measured.

Here we are weighing groups [28]

Skipper and Cook believed that women couldn't run hydraulics on a boat because we were incapable of operating such intricate devices. During the first couple of voyages, I carefully watched staff and crew operate the hydraulics located on the deck. When Skipper wasn't looking, I asked the First Mate to explain the hydraulic system. One day, when it came time to power up the hydraulic winch, I quietly stepped behind the controls and completed the task with no problem. All the guys stopped dead in their tracks to stare at me, but only for a second, then they went about their jobs with no comment. Nothing was said, and from then on, I assumed my place behind the controls when it was my turn to help.

Cook hounded me relentlessly. On the fall 1984 trawl survey, he told me, "I'd like to stick my tongue down your throat." When I cautioned him to tone his remarks down, he became angry. I often had to deal with crude language. When the shellfish manager called me a bitch, and I said I didn't want to hear that kind of language, he replied, "I don't give a shit." I suspected that the men who made these remarks sought power by treating me disrespectfully.

The May 1984 estimated abundance of shrimp in Kachemak Bay ranged from 3.0 to 5.2 million pounds.[35] Because catches varied greatly by station, from zero to a full net, it affected precision of the estimate. The October 1984 estimated abundance was a whopping 6.1 to 9.1 million pounds.[36] As expected, pink shrimp comprised the greatest percentage of the catch in both spring and fall surveys. Sidestripe shrimp were the next most common species, followed by humpy shrimp. Generally, humpy shrimp are only available to trawl gear in the fall because they move to waters outside usual trawl areas at other times of the year.

The 1985 surveys found fewer shrimp than in 1984. Based on our surveys, the shellfish manager set the 1984 commercial harvest limit to 1.6 million pounds, and in 1985, he set the harvest limit to 1.2 million pounds. Clearly, the abundance of shrimp in the bay was decreasing.

The incidence of fish in the tows increased over time in waters west of Homer Spit, and I was concerned about that because walleye pollock, Pacific cod, flathead sole, and arrowtooth flounder are predators of shrimp and crab. A climatic regime shift was beginning to occur across the northern Gulf of Alaska, and ocean temperature changes were altering fish distribution. The consequence of large predator populations moving into Kachemak Bay could lead to decreasing shrimp biomass. In subsequent years, this was indeed the case. By the late 1980s, shrimp biomass had dwindled to such an extent that the commercial trawl fishery in Kachemak Bay ended. Reasons for the decline in shrimp abundance may be a combination of overfishing, increased predators, and ocean temperature changes.

King and Tanner Crabs

The earliest commercial landing of king crabs in LCI occurred in 1937. Harvest was low in the 1940s, but by the mid-1950s had risen to 2 million pounds a year, with most of the harvest in the Southern District. During the early 1960s, fishing for king crabs expanded into Kamishak and Barren Islands Districts and boats harvested up to 8 million pounds a year.

In the early years of king crab fishing, Tanner crabs were an annoying incidental bycatch, but in the late 1960s, demand and price for Tanner crabs increased to the point that fishermen began to target Tanner crabs in all five districts of LCI. For a while, both the king and Tanner crab fisheries chugged happily along. In the late 1960s through mid-1970s, the annual catch of king crabs ranged from 2.5 to 4.8 million pounds, and for Tanner crabs ranged from 1.4 to 8 million pounds.

Then, the harvest of king crabs fell dramatically. The Southern District king crab fishery was closed in 1982, and the Kamishak District fishery was closed in 1983. The local crab fleet relied on Tanner crabs as a source of income for their families while they waited (in vain) for king crab stocks to recover.

The early management strategy for crab stocks was to allow only males of a certain size to be harvested. Over time, size limits were raised to enable males more time to mature and mate before the fishery targeted them. Additional restrictions, such as guideline harvest levels, season closures, and pot limits, were implemented to curtail harvest. In the case of king crabs, these measures may have been like closing the barn door after the cows had already escaped.

Initially, research of king and Tanner crabs in LCI entailed simply sampling the commercial catch for biological data while talking with fishermen. In 1974, a cooperative state-federal research program for king and Tanner crabs was conceived to estimate relative abundance and size, sex, and age composition in an index survey using commercial pots. The state assumed responsibility for the surveys in 1982. The shellfish manager used relative abundance indices from the surveys for both species to set guideline harvest levels in the commercial fisheries. My job was to continue the surveys as they had been previously conducted. The Kamishak District survey occurred in June, closely followed by the Southern District survey in July.

To prepare for my first crab survey, my assistant and I went to the storage yard on the spit to inspect the pots before transporting them to the dock. Our pots were the same commercial-sized pots as had been used since 1974. They measured 7 feet by 7 feet with a 3.5-inch stretch mesh. Each pot weighed 700 pounds and cost my limited research budget $800 each. Loading king crab pots onto a trailer for transport to the dock was a new chore for us. The crab technician, Bubba, refused to help, even though he was experienced with the process. He remarked, "I just don't see why working for a woman is necessary," and stalked off. His opinions interfered with state business at times. My assistant and I figured out how to load the pots without advice from Bubba.

We drove the heavily-loaded trailer to the dock; the task of off-loading the pots onto the boat was left to an experienced crane operator. The operator worked carefully to stack pots for even weight distribution and to avoid dropping a heavy pot onto the deck.

Once loaded, the pots were lashed tight because if the load shifted, the imbalance could capsize the boat.

After the pots were loaded onto the stern of the *Pandalus*, we motored out to sea. In addition to myself, the crew consisted of Skipper, First Mate, Cook, my assistant, a technician, and a finfish biologist. Fish (particularly Pacific halibut and Pacific cod) were common incidental captures in crab pots, and the job of the finfish biologist was to identify, count, and measure the fish. Pacific cod, voracious predators, were periodically dissected to check their stomachs for shrimp and juvenile crabs. I routinely saw Pacific cod with as many as twenty-five juvenile crabs in their stomachs.

That day in June 1984, the water was still, the sun bright, and the breeze warm. We had a fair distance to reach the first of forty stations selected from a grid pattern overlain on a map of Kamishak Bay. Skipper used a Loran C radio navigation system to pilot us to the same grid pattern as had been sampled for the past ten years. The grid pattern had been selected in the early days based on crab catch concentrations reported by commercial fishers. Each station was 5 nautical miles^2 in size. Within each square station, five pots were placed in a line, evenly spaced about 0.3 nm apart. The direction of the line of pots was determined by bottom topography, tidal flow, and the most efficient vessel course. Yellow rope attached a bright, pink buoy to each pot, so we could easily see the pot and hook the buoy for pot retrieval.

As we approached the first station, a pot was plucked from the stack with the winch and guided to settle on the pot launcher. Chopped frozen herring was placed in perforated 2-quart bait jars to hang in each pot. We gauged the length of rope needed to float the buoy based on water depth and tidal action and coiled the line at the base of the pot, ready for launch. The buoy would sink below the surface if the rope was too short. When Skipper signaled that we were at a station, the pot was tipped over the side into the sea. The rope and buoy were thrown after it. Four pots followed in succession to complete the first station. After all the pots were deployed, we anchored and waited twenty-four hours, the standard soak time.

The next morning, the sea undulated in long swells. Working with the heavy pots became challenging. We were on guard and alert. As Skipper eased the boat past the first buoy in the string of pots, I threw out a hook to snag it. As the buoy and rope were retrieved, the rope was slipped into a winch that pulled the pot starboard.

A crew member swiftly coiled the rope on deck. As the pot was raised out of the water, it was guided to rest on the launcher. The pot's panel was unhooked, and the entire catch was dumped into a big tote. While we quickly sampled the catch and recorded information, the empty pot was maneuvered onto the stern to make way for the next pot in the string. We worked in synchrony, quickly and carefully, to keep the process moving.

Crabs were sorted by species and sex, counted, measured, and assigned to age groups. For example, "legal" king crabs are males greater than 7 inches in carapace width. "Recruits" are new-shell males that have molted in the current year and are 5.5 to 7 inches in size. "Pre-recruit 1" king crabs are males 5 to 5.5 inches in carapace width and denote the number of molts required to reach legal size.

The egg condition and clutch size of mature female king crabs were estimated and periodically sampled. Females carry from 40,000 to 500,000 fertilized eggs beneath their abdominal flap for eleven months before larvae are released to the currents. Following sampling, crabs and fish were returned to the sea. We wore thick gloves to protect our hands from the sharp red spines of the king crab's exoskeleton and were careful to avoid their claws, which had sufficient strength to shear off a finger. King crabs are big and powerful. They can live up to thirty years, with females weighing 10 pounds and large males weighing 20 pounds.

Collecting the crab catch [28]

During our crab survey in Kamishak District, sea conditions ranged from calm to rough. On a calm day, Skipper dropped anchor off Augustine Island and let us take the skiff to explore the sandy beach for flotsam. This small and uninhabited island was formed by a volcano that is still active. The snow-capped dome rises to 4,100 feet; periodic eruptions deposit ash and lava debris that makes up most of the landscape. Plants and animals have colonized the edges of the island. Grasses, sedges, and lupine extend from the beach to the volcano's base, and shrubby alders cling to its lower slopes. The island is an important habitat for sea and shore birds. I don't know how foxes initially arrived at the island, but we found a thriving fox population that survived on birds by hunting them, stealing their eggs, or scavenging the dead.

On another day, strong winds kicked up. Skipper said a typhoon was forming, and we needed to seek shelter to ride out the storm. Surging through high waves, the *Pandalus* nosed into Iniskin Bay on the western side of Cook Inlet, and we lay at anchor the rest of that day and through the night. Rocky cliffs around the bay protected us from the full force of the wind gusts.

In 1984, a total of 185 pots were pulled from forty stations, capturing 279 male king crabs clustered in two areas. The 1984 mean catch-per-pot of legal male king crabs was 0.4, the lowest on record. The 1985 survey results were not much better. The surveys suggested the abundance of legal male king crabs was low, and the trifling catch of pre-recruit 1 male crab painted a dismal picture for the future as it was a tenth of the previously recorded lowest catch.[37] Like the males, the female mean catch in both 1984 and 1985 was the lowest on record. Because of the low catches of king crab, continued closure of this fishery was recommended in both 1984 and 1985.

The total number of legal male Tanner crabs in the Kamishak District surveys in 1984 and 1985 was 771 and 781, and the mean catch-per-pot was 4.2 and 4.3, respectively. The catches were higher than the previous two years, during which commercial fishing seasons were allowed, so I expected the manager to proceed with Tanner crab fishery openings. The commercial harvest of Tanner crab in 1984 and 1985 was 1.54 million pounds and 1.29 million pounds, respectively.

A troubling incident occurred toward the end of the Kamishak crab survey. Before our departure, Al had dropped by my office to offer his advice. His last words of warning to me were, "Don't let the guys turn this into a meat run." I wasn't clear about his meaning at the time, but I gained insight. One of the staff began to throw live king crabs into an industrial-sized tote containing ice instead of returning them to the sea per protocol. It became apparent he was harvesting his winter's supply of king crab, which was illegal and unethical. Because the survey was under my supervision, I told him to stop. In response, he called me a bitch, said he would not stop, and made this threat: "Don't turn your back on me!"

Honestly, I feared for my safety after that. Crab fishing in Alaska is one of the world's most dangerous jobs for injuries and deaths. Dangers included: getting tangled in rope as a pot is launched overboard, thus carrying you swiftly to the bottom of the sea; losing your balance as you attach bait jars, thus falling into the pot and knocking it overboard, carrying you to the bottom of the sea; being crushed by a pot; or, being swept overboard during high seas. Any of these "accidents" could have been arranged. Nonetheless, I raised holy hell with Skipper. He got so tired of my complaints that he made a new rule: in the future, staff and crew could only keep incidental mortalities that would fit in a small bag.

Upon reaching the Homer dock, the staff member used the dock's hydraulic crane to hoist his tote filled with illegal crabs from the deck of the *Pandalus*. I could see the accusing glances from workers on the dock—they knew what was going on. No wonder the snide nickname for the state vessel among fishermen was the *Scandalus*!

One week after returning from the Kamishak District crab survey, we went back on the water for the Southern District crab survey. A new survey and a fresh start with the crew. Chuck had repeatedly encouraged me to "create a team atmosphere." Accordingly, I laid the past aside, focused on the present, and carried on.

The Southern District survey covers more area; usually, 230 pots are pulled from sixty stations. The mean catch-per-pot of legal male king crabs was 1.8 in 1984 and 1.2 in 1985, which was below average. Like findings in the Kamishak District, pre-recruit 1 male king crabs were very low in abundance.[38] Based on our survey, the shellfish manager kept the king crab commercial fishery closure in place. This commercial fishery has never re-opened due to the persistent low abundance of king crabs. We thought diminishing crab populations were likely due to a combination of changing ocean conditions, increasing numbers of predator fish, and overfishing.

In contrast with the king crab survey results, the catch of Tanner crabs was amazingly high, rivaling high catches of the 1970s. The mean catch-per-pot of legal male Tanner crabs in 1984 and 1985 was 24.9 and 35.4, respectively.[38] News of the high survey catches caught the eye of nonlocal fishermen, much to the ire of the Homer fleet. In 1984, eighty-three vessels harvested 1.23 million pounds of Tanner crabs from the Southern District. In 1985, 103 vessels fished for Tanner crabs, harvesting 1.16 million pounds. The Tanner crab commercial fishery in LCI lasted another ten years before it was permanently closed in 1995 due to low crab abundance.

Cook continued his sexist remarks on the crab surveys, stating I had a "nice fanny." Once, when I was lying down between stations, he jumped on top of me. I pushed him off and told him to cut it out. Cook wasn't threatening; he was annoying. I tried different strategies with Cook: I ignored him, laughed at his remarks with the others, warned him, got angry with him, and complained to Skipper about him. I tried every tactic I could think of to thwart his unwanted behavior, but he wouldn't stop. At that time, I didn't file a formal grievance out of concern I would be perceived as a hysterical woman and consequently blacklisted by the guys in charge from future employment opportunities. I sought to be perceived as a gender-neutral biologist, judged on my professional merits. Grieving harassment would have shined a spotlight on a situation I was striving to avoid, but I learned that avoidance postpones the inevitable reckoning. I think today's woman is less apprehensive about the fallout from exposing unacceptable conduct than I was in the mid-1980s—at least, this is my hope.

Cook and Skipper were at odds with each other because Cook used to be a captain and had been demoted. They always put each other down, and it wasn't easy for them to work together. Skipper knew that Cook had a short temper and deliberately irritated him to enjoy watching him fly off the handle. How could I create a "team atmosphere" with those two? It was a personnel manager's nightmare!

Chuck, recognizing that personnel conflict in the Homer office made teamwork challenging, sent me to a conflict management seminar in Anchorage. Perhaps he hoped the training would give me insight into solving personnel conflicts in the Homer office that Anchorage supervisors had thus far failed to deal with. Chuck wrote on my evaluation, "Peggy responds well to supervision from the regional office. Interpersonal conflicts have occurred with other staff members within the Homer office. Such conflicts have been a frequent problem in the Homer office before her arrival and continue to exist. Peggy recognizes these problems and strives to maintain good communication with several very independent-thinking biologists. Getting along with all existing staff will require extra effort on her part and considerable tact." It seems that it was my responsibility to be the peacemaker.

I was usually upbeat and didn't let the intransigence and harassment of the guys in the Homer office get to me, but one day I felt down. I thought it would be helpful to commiserate with another female biologist in the Commercial Fisheries Division who was also stationed in a field office and had area responsibility. I sought support to help me regain a sense of control over my work life and hopefully learn workaround tactics to achieve my professional goals. I thumbed through the staff directory. The only other woman I found was named DeeDee, who had worked in the Bethel area as a management biologist until she quit to become a world-renowned dog musher. I'd never met her, but I called and asked about her experiences working as a woman in ADFG and if she had any advice. She kindly listened and offered support. Articulating my frustrations out loud was good therapy. I appreciated talking with someone who understood what it was like to be a woman in a traditionally male-dominated profession.

The history of professional women's employment in ADFG up to 1985 was dismal. After DeeDee quit, I was one of just four women who were stationed in a field office with responsibility for area research or management in the Commercial Fisheries, Sport Fish, or Game Divisions. Women were not being promoted to high ranks at the rate that men were. In 1985, the employment status of professional women was below target; while women at range 18 or higher comprised 8.6 percent of ADFG's workforce, the target set by ADFG was 24.7 percent.[39]

Parasites

In 1983–1984, the abundance of king crabs in half of Alaska's management areas suddenly declined. One theory for the decline was reduced fecundity (number of eggs) in females. So, in the early 1980s, fecundity, represented by "percent of egg clutch fullness," was noted when conducting surveys.

In collecting fecundity data on the crab surveys, I sometimes found eggs that looked gray and gooey, like they were diseased or dead. This raised suspicion of a parasitic worm that could be causing egg mortality. After discussing a possible egg parasite with a Kodiak shellfish biologist, Forrest, and a researcher at the University of Alaska's Seward Marine Center Laboratory, A.J., I expanded the collection and examination of crab eggs through space and time. To determine the incidence of suspected parasitic egg infestation in king crab eggs across Kachemak Bay, I took a systematic sample of eggs during our surveys. Crab eggs were also sampled during the Kamishak Bay surveys.

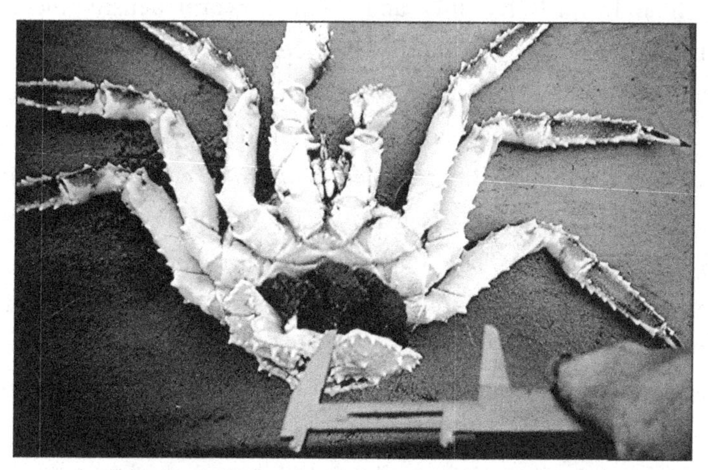

I find a full egg clutch in a female king crab

To obtain information on the prevalence of the suspected parasite over time, I collected king crab egg samples periodically from one location in Kachemak Bay in 1984 and 1985. The samples were slipped into vials of preservative and shipped for analysis to A.J. and a parasite specialist at the University of California, who was on contract with ADFG. Working with these specialists, we determined the suspected parasite's role in egg death and evaluated its presence and impact on the crab populations in LCI.

A new species of nemertean worm predator, measuring 1 mm in length, was found in the Kachemak Bay samples! Worm larvae float in the ocean and find crabs to settle on, where they encyst in the crab's exoskeleton. Worms on females migrate to the egg clutch when females release eggs under the abdominal flap. Worms on males can transfer to females during mating. The worm preys on crab eggs by using a stylet to puncture the egg casing, then everting its esophagus into the opening to ingest the contents of the egg.[40,41]

Worm density in eggs was low shortly after they were laid in March and April. By July and August, worm density increased sharply, with egg mortality between 45–85 percent. Worm prevalence and egg mortality were high in September and October, and egg losses approached 100 percent. The systematic sampling scheme in Kachemak Bay revealed that female crabs throughout the bay area were affected.[41] Egg loss was virtually total in the 1983–1984 brood season. The intensity of worm infestation and egg mortality subsided slightly in the 1985–1986 brood season. Interestingly, while Kachemak Bay king crabs were severely affected, the worm was less prevalent in Kamishak Bay king crabs.

We discovered that our attempts to estimate fecundity by classifying the percent of egg clutch fullness using observation were inaccurate because we couldn't always see dead eggs. For example, in June 1984, nearly all female king crabs sampled were classified as having egg clutches at 90–100 percent full. However, later lab analysis revealed that dead eggs comprised up to half of some samples.[41]

The cause of the sudden appearance of the parasitic worm infestation was unknown. The worms have likely always been in the ocean at low levels, but why they exploded into an epidemic in certain crab populations remains a mystery. Perhaps worm outbreaks have periodically occurred in crab populations in the past but went unnoticed. The brood failures from near total egg mortality reduced the number of crabs available for harvest and prolonged recovery of crab populations decimated by other factors, such as overfishing. The worm explosion and resulting egg predation in 1984 and 1985 reduced the reproductive output of king crabs in Kachemak Bay. All the manager could do was close the fishery and hope the crab population recovered.

With the discovery of a new parasitic nemertean worm in king crabs, great interest was kindled in looking for similar pathogens in other Alaskan shellfish. I began to receive requests for tissue samples from researchers with ADFG in Kodiak, the University of Alaska, and a pathologist specializing in shellfish diseases and parasites at the National Marine Fisheries Service in Seattle. I agreed to assist with their research, so they sent me histopathology protocols and sampling equipment. These extra tasks took additional time, but the research was important. In 1985-1986, a dinoflagellate disease afflicted Tanner crabs in some areas of Alaska, making the meat bitter-tasting and unmarketable. Dinoflagellate-like parasites were also discovered in Alaskan spot and pink shrimp—the infection caused shrimp meat to turn chalky. Fortunately, these dinoflagellates were not detected in the Kachemak Bay crab and shrimp samples I collected in 1984 and 1985.

Tanner Crab Pot Soak Study

In the 1970s–1980s, ADFG established annual harvest guidelines for Tanner crabs based partly on index surveys. Survey pots were normally fished for twenty-four hours. However, weather conditions and mechanical problems sometimes delay the pot's retrieval, resulting in longer soak times. Because crab catch-per-pot changes with soak time, catches should be adjusted to reflect a standard soak time and thus avoid biased estimates of mean catch-per-pot. ADFG treated Tanner crab catch data with nonstandard soak times differently between management districts. For example, staff in the Kodiak district devised correction factors from commercial logbook information, whereas, in Kachemak Bay, no correction factor for soak time had been used. I didn't know how to treat the data.

I needed a method for adjusting catches based on a model describing the change in the number of crabs within a pot over time. This issue was one of many tumbling around in my mind when I chanced to meet a biometrician at a scientific conference who brought up this issue. David worked at the National Marine Fisheries Service in Seattle, and while he had some modeling concepts in mind, he didn't have a way to field test them. I offered to join forces with him to conduct an experiment on soak times applied to Tanner crab. David and I collaborated on the study design, and in early 1985, I received permission to proceed with the research. Ours was among the earliest studies to examine the relationship between Tanner crab pot catch and soak time.

I told the Homer staff and the *Pandalus* crew about the intended research. My idea was met with resistance. The staff and crew were suspicious about anything they were not accustomed to, and the soak study was a new concept. Bubba refused to participate in the study. A week later, Cook stormed into my office and told me that he'd talked to ten fishermen, and they agreed my soak study was a waste of money. I invited him to sit down and tried to calmly explain the study. But I warned him that research decisions were my job, not his, and I did not appreciate his efforts to undermine public confidence in state research by grumbling about it on the docks. If fishermen had questions about the study, they could come to see me.

When we motored out to set the pots, the crew's reluctance to participate in the research came to a head. On day 1 of the study, my journal reads, "Got shit from crew again. They are fighting me because they don't see concepts. The crew is also fighting me on the timing of the study. Conflicts that must be worked around include the crew's bowling schedule. Cook said he hoped the soak study failed. Their bad attitude hinders my job."

We set out three types of pots: traditional pots, ones with modified entrance tunnels to allow crabs to enter but not to escape (entry-only), and pots that allowed crabs to exit but not enter (exit-only). All pots were fished the same, except legal-sized male Tanner crabs were put into the exit-only pots at the beginning of the study. Pots were retrieved after a twenty-four-hour soak, and crabs were counted but not removed. A marking scheme was devised to identify crabs present in the pot per day, so that newcomers would be readily apparent. Pots were reset in their original positions without new bait.

On day 2 of the study, my journal reads, "Cook made a lot of sexist remarks about keeping women in their place. They belonged in a kitchen, not on a boat. I thought that was rich, coming from a male Cook! When I asked him to handle crabs differently, he said he wouldn't handle them at all. I asked Skipper to speak to him about cooperation." After the first two days of the study, the crew got used to it, and the remaining days went smoothly, although, on the last day, Cook lobbied a final salvo about the study, telling me, "You have your head up your ass."

I complimented Skipper on a good job. He just told me to clean the deck, whereupon I reminded him that turnabout is fair play—I'll help clean the boat if the crew helps us mend gear. We left it at a standoff. In a gesture of peace, I created a comprehensive trauma kit for the *Pandalus* by buying medical supplies from my research budget and presented the kit to Skipper. The kit was intended to replace the inadequate tin of band-aids and aspirin the boat carried. Given the serious risks of our work at sea, we needed more advanced treatment options. I think Skipper appreciated the medical kit; he offered to replace my costs with $230 worth of rope.

Back in the office, I organized the data and sent it to David in Seattle. We had a lot to think about. Common sense suggests that the number of crabs in a pot will initially increase as they seek the bait, but their rate of increase will decline with time as the local density of crabs, as well as bait effectiveness, is reduced. Developing a model to adjust catch-per-pot was complex because we had to consider the effects of crab density on entry rate or escape probability, social behavior (such as if crabs are attracted or repulsed by other crabs), and the declining effectiveness of the bait. Also, environmental variables, such as water temperature, influence not only the crab's search speed but also the bait's effectiveness and, thus, entry and exit rates.

David designed a model to estimate mean catch-per-pot, and we anxiously pored over the tests. The model suggested that the current (1985) methods used by ADFG to treat Tanner crab catch data with nonstandard soak times (whether no correction factor was used or the Kodiak correction factors were used) underestimated the catch. The degree of underestimation progressively increased with greater soak times. Our results were published in a professional journal,[42] and in subsequent years, others cited our small contribution to the research of crab pot catch and soak time.

Dungeness Crab

Dungeness crabs were harvested in the bays of the Southern District beginning in the early 1900s. While some landings occurred in the

Outer and Eastern Districts, fishing Gulf of Alaska and Cook Inlet waters was less productive than bay fishing and offered greater hazards. In 1973, a large stock of Dungeness crabs was located off Bluff Point, 5 miles northwest of Homer. Most of the fishing effort shifted to Bluff Point in the early 1980s.[43]

When I arrived in Homer in 1984, there was a thriving Dungeness crab commercial fishery. The mean ex-vessel value of the fishery was $440,100 per year, and the mean catch was estimated at 615,200 pounds, with the highest reported at 2.1 million pounds in 1979. With the decline in king crab fishing, many fishermen had switched to harvesting Dungeness crabs.

Harvest monitoring of the Dungeness crab fishery got underway in 1964. Early regulations allowed a twelve-month season, but fishermen generally confined harvest to summer because storms and ice severely restricted winter harvest. Management of the fishery evolved to a 3-S concept: sex, size, and season. Only males with a carapace width of at least 6.5 inches could be harvested.

No index surveys were conducted, but there was a dockside sampling program. During the fishing season, I spent time at the Homer dock when the boats came in and measured a sample of their catch. Size of the crabs helped evaluate strength of a year-class. Observations on disease or soft-shell conditions helped gauge population health. Dockside sampling also provided an opportunity for me to talk with fishermen. The catch-per-unit-of-effort in the commercial fishery, total catch landed, and dockside sampling were used to roughly assess the status of the population.

Early research aimed to trace crab distribution and understand their biology through a tagging program. Between other tasks, I pulled out the raw data on an old tagging program to review and re-analyze it. I used a bivariate covariance matrix to estimate the area covered by tagged and recaptured crabs. The model indicated that area covered by crabs differed between geographic regions and sexes. For example, crabs tagged in Port Graham and Seldovia Bay tended to remain in those bays. Crabs tagged along McKeon Flats migrated extensively along the coast. Bluff Point crabs migrated within a 120 to 130 square-mile area off Bluff Point, and their movement did not overlap with crabs tagged along the coast between Port Graham and China Poot or in upper Kachemak Bay.

Non-overlapping movements of adults tagged in different areas suggested they were reproductively isolated. While larval drift movements must be considered before suggesting the existence of separate stocks, currents in Kachemak Bay consist of stable and predictable gyres, increasing the localization of larval settlement. Adding to the implication of separate stocks, catch sampling found variations in the size and shape of legal male crabs from different areas. These variations may have been more environmentally induced than genetic but pointed to differences in growth conditions.

I presented my findings at a Dungeness crab symposium sponsored by Alaska Sea Grant in October 1984.[44] A possibility of separate Dungeness stocks in LCI raised some interest. The concept of Dungeness crab stocks along the same coastline was not novel. California Fish and Game had identified Dungeness crab stocks and managed each with different fishing seasons based on crab shell conditions. Shell condition can be a significant limiting factor in the marketable catch. Processors do not purchase soft-shell crabs. Fishermen stated their catches of soft-shell legal males in the Southern District varied from 5 to 90 percent, with the percentage of hard-shell legal males increasing (but not consistently) over the season.

Maybe variability in soft-shell condition in legal male Dungeness crabs in the Southern District was tied to microhabitat and could be predicted by area, but I didn't know enough yet. I did know that fishermen were eager to find a way to avoid catching soft-shell crabs so they could reduce loss from handling mortality and deliver a product they could sell. They hoped my research would help clarify the matter.

In February 1985, Garland Blanchard came by my office to say fishermen were encountering soft-shell crabs, and processors wouldn't buy them. Garland wanted the fishing season closed through May. There was no use in harvesting crabs they couldn't sell—it was an economic waste! Although Garland knew I was not in charge of managing the fishery, he hoped I would have some influence on issuing an Emergency Order. However, I was ignored by the shellfish manager as much as the commercial fishermen were, so nothing changed. During the rest of the winter and spring, other fishermen came to see me, all complaining about the unusual number of soft-shelled crabs they were catching and their difficulty in selling them. Finally, on May 9, an exasperated buyer with Seward Fisheries, Rob, called me and, in no uncertain terms, said he wasn't going to buy Dungeness crabs until June 1. He wanted the Dungeness fishing season in the Southern District closed, by God, because he was tired of getting complaints from fishermen!

Where was the shellfish manager during all this? Why did the fishermen and processors turn to me for help? Probably because they knew I cared about what they had to say. But, given the hierarchy and culture of the Homer office, my powers to enact change were minimal. Nonetheless, I attended meetings with commercial fishermen at Seward Fisheries, organized by Rob, and answered their concerns and questions as best I could. Rob continued to call me with updates on the quality of crabs coming in through the fishing season.

Unfortunately, the regional supervisor decided to reorganize LCI staff. He changed my boss, replaced the old shellfish manager in Homer with his counterpart from Cordova, and gave my assistant to this new shellfish manager. The regional supervisor intended to give the Fishery Biologist III position that had belonged to my predecessor and that Chuck had promised me to the new shellfish manager. In fact, on one of his visits to Homer, the regional supervisor told me, "You'd have to act like a Fishery Biologist IV for me even to consider elevating you to a Fishery Biologist III!"

The message given to me was that I must do the work of a Fishery Biologist III but be paid at a Fishery Biologist II scale; if I wanted to be considered for a promotion, I would have to perform two levels above my pay grade.

With the arrival of the new shellfish manager, animosity toward a female biologist in the office rose. Playboy centerfolds appeared on the bathroom walls. My former assistant posted a photo of a nude woman on the front of his door, where the public could walk right by it. I overheard Skipper tell the crew that he heard the new shellfish manager on the radio of the Cordova research vessel, the *Montague*, saying, "There is no way I'm going to let a woman work on my boat, and I sure hope a women's group doesn't hear about it."

Shortly after that, I was summoned to Anchorage by my new boss, Dick. I got into my Bronco in the early morning and drove three and a half hours north. After walking into his office and sitting down, he told me that Skipper had passed along a complaint Cook had made about me. I asked, "What is Cook's complaint?" Dick replied, "You left your coffee cup on the galley table."

I stared at my new boss for what seemed like a long time in silence. He had required me to make a seven-hour round-trip drive so he could bring up a trivial offense—what kind of game was he playing? Well, I wasn't going to be intimidated into taking a defensive stance. I got up and walked out. There was no point in talking to someone who was deliberately blind to my offensive work environment in Homer or, worse, complicit. The situation bordered on farcical. I drove back to Homer and, the next day, went to work as usual.

Except it wasn't a usual day, it was a really bad day. I received news that the only two children of my friend and EMT partner in Glennallen, Peri McIlroy, had been hit by a drunk driver. Kurt died instantly, while twenty-year-old Kara lived five more days before dying on August 25, 1985. I couldn't imagine Peri's anguish. When I called her, tears choked my words as I promised to find a meaningful way to honor her children's memories. In their names, I am a life-long supporter of Mothers Against Drunk Driving, an organization that advocates for victims and advances measures to reduce drunk driving. Kurt and Kara were buried in the Glennallen Cemetery, and Peri was laid to rest beside them in 2012.

On September 20, the phone in my office rang, and when I picked it up, the Chief Fisheries Scientist, Phil, said, "We have to get you out of there. Come work for me." Phil explained that he was creating a statewide shellfish research position stationed in Anchorage under the supervision of a Biometrician II. While I would be transferred as a Fishery Biologist II, he wanted me to qualify for the Biometrician I register. I would need to take courses in statistics at my own expense and on my own time. I was surprised to be asked to join his team and requested time to think about it.

I loved my job in Homer. It was meaningful—people's livelihoods depended on my research. There was much to discover about shellfish biology, and I felt I could contribute to the science. Homer was a great place to live and enjoy outdoor activities. Finally, I valued my time with the Homer Fire Department and my friendships there.

However, I didn't see any hope of being accepted by the Commercial Fisheries Division staff in Homer and working together as a team. I was tired of having to persevere through the harassment. The deputy director called and encouraged me to accept the job offer. So, I did. The position in Anchorage would begin on January 1, 1986. I had three months to tie up loose ends in Homer. At the end of November, the new shellfish manager asked when I was leaving. He said he hoped it would be soon, so he could "get on with things."

I left Homer with no regrets. The job had been a study of human behavior— men resisting change to their preconceived ideas of what a stable workplace looks like. Resistance to change continues today in many parts of society. Hopefully, not so much at ADFG.

Life in Homer

The Anchor River, a fifteen-minute drive north from the office, is a pretty 30-mile stream draining into Cook Inlet that supports runs of salmon and steelhead. In summer, I met friends in the early morning and fished the Anchor River before going to work. Sometimes, we drove farther north to the Ninilchik River and Deep Creek to go fishing. Off Anchor Point, we trolled for halibut. If you hooked a big one, it felt like reeling in a sheet of plywood—it was hard work to haul the halibut into the skiff, and you didn't want to be in the way of that heavy tail flapping.

The Kenai River, famous for record Chinook salmon, trophy rainbow trout, and Dolly Varden, was a 1.5-hour drive north of Homer. In summer, its waters, milky turquoise in color from glacial silt, were crowded with anglers. A few times, I joined the throng. I'd get up early on a weekday to be on the water at daybreak, which in summer was around 4:30 a.m. I fished salmon eggs soaked in a secret solution using a sturdy bait-casting rod for a few hours, then drove back to arrive at the office for work on time.

One day in 1986, a friend at the office, Nick, met me on the Kenai River, and we hired a guide to take us out on the water at dawn. After several casts, my lure got stuck. I was so annoyed! The guide positioned the boat so I could break free from the snag, but the lure wouldn't budge. Then, a huge force ran my line downriver, almost toppling me over the side of the boat. The lure wasn't stuck. I had hooked a big fish! For an hour, I reeled in the line, then let it race out as the guide chased the salmon. My back ached, and my arms felt tired, but I slowly drew the fish to the side of the boat. Our jaws dropped when we saw its size. I had landed an 80-pound salmon! That morning, I got to work late.

My 80-pound Chinook salmon from the Kenai River [45]

While most of my outdoor recreation was fishing, I enjoyed duck hunting and shooting at the Homer Trap Range, where I met my future husband, Jim. One day, Jim and I were shuttled to the head of Kachemak Bay for a weekend of duck hunting. The sky was overcast as we anchored decoys in a small pond and set up our tent. At sunset, a howling wind poured down from Nuka Glacier through Bradley Pass. Its strength and speed caught us by surprise, flattening our tent and forcing us to seek shelter in an old cabin. Half of the cabin's roof had caved in, and the floor had sunk at one end, but we hunkered down in a corner and waited for daylight. At dawn, the wind abated, and we awoke to find our decoys trapped in a frozen pond. We watched ducks fly high overhead in the clear blue sky, out of range, seeming to laugh at us standing next to our pathetic frozen decoys. After that first night, the rest of the trip went smoothly, and we got a couple of ducks for dinner.

***

Shortly after arriving in Homer, I joined the Volunteer Fire Department. I became a Cardio-Pulmonary Resuscitation instructor and completed an EMT II course. In Alaska, the EMT II is licensed to perform venipuncture—inserting a needle for intravenous therapy. Under the direction of a doctor, EMT IIs can administer certain medications. For my final exam, I put a line in myself. I used a vein in my foot and set up an IV saline drip. I enjoyed the camaraderie of the fire department crew, all good and dedicated people, and our shared interest in offering medical aid to the community. The emergency calls I responded to in Homer were generally routine, except for three that I clearly remember.

I had Chuck's permission to leave work if I was summoned for a medical emergency. There were only a few occasions when that happened, one of which occurred on August 16, 1984. I arrived at the scene of a horrific car accident. A young woman was so intertwined with the crumbled metal of her car that we couldn't see or access her lower body. The fire crew operating the Jaws of Life had a hell of a time carefully peeling off the layers of metal from where we presumed her lower body was. She was unconscious, but we detected a faint pulse. My job was to secure the airway and stabilize the cervical spine. I reached for her head through what was left of the back window, contorting my body around the twisted metal. She had an unobstructed airway. There wasn't enough space to apply a cervical collar to stabilize her neck, so I held both sides of her head steady with a gentle pull and stayed immobile until she was extricated. My muscles cramped, but I held steady as her life could depend upon it. I made sure her neck did not move as the layers of metal were cut and lifted off. Other ambulance crew members controlled her bleeding and got a line in her arm. Finally, her body was freed from the wreckage. More EMTs swarmed in to stabilize fractures and bleeding. We carefully lifted her onto a gurney, put on that cervical collar, and loaded her into the ambulance. I remember how fretful I was as I paused to look at her on the gurney. I was not sure she would make it. If she did survive, she would have a long rehabilitation ahead of her. With a heavy and worried heart, I returned to the office and tried to resume work.

I was on call when the Homer Police Department requested an EMT to accompany an officer to a domestic disturbance. A woman was reported acting erratically by her landlord. When we arrived at the apartment, the officer knocked on the door and announced who he was. The woman who opened the door was stark naked and appeared to be disorientated, so I stepped in, talked to her, assessed there were no life-threatening conditions and helped her find a robe to cover herself. We decided to transport her to the hospital for an evaluation. The officer eased the woman into the back of his patrol car and asked me to sit beside her to keep her calm. Big mistake! She was unrestrained, and she became increasingly agitated. Suddenly, she lunged at me and bit me on the arm. Ouch, that hurt! Her teeth broke my skin. In 1984, HIV infection was beginning to be reported, all emergency personnel knew about its presence in Alaska, and I was worried about her infecting me. Thankfully, the test results came back negative. I learned not to sit in the back of a patrol car with an unrestrained mental patient!

The third call that sticks in my mind occurred on a snowy night. I was called to the helicopter pad. An emergency message had come in from one of the Russian Old Believer communities at the head of Kachemak Bay, Kachemak Selo. A two-year-old child had stuck her finger in an electrical socket and had been electrocuted. There is no road to this community, thus the need for helicopter transport. The snow was coming down thick. Five of us were assembled at the helicopter pad: the pilot, two EMTs, the EMS Chief, and his assistant. There was a question about the weather. Was it safe to fly? After a few minutes of discussion, the EMS Chief raised his hand and said, "No. I'm not going to risk the lives of personnel in this snowstorm for a child who is likely already dead." We were sad, as we were willing to try to reach the child, but the EMS Chief made the right call. I didn't envy his somber responsibility for making life-and-death decisions.

My time spent with the Homer Volunteer Fire Department was my last opportunity to volunteer as an EMT. After leaving Homer, I was stationed in towns large enough to support salaried EMS staff, and became more involved with work and my private life.

Statewide Shellfish Research

"Every time you meet a situation you think at the time it is an impossibility and you go through the tortures of the damned, once you have met it and lived through it, you find that forever after you are freer than you were before."
~ Eleanor Roosevelt

I moved to Anchorage, rented a small apartment, and enrolled in statistics courses at the University of Alaska. Most of my free time was spent doing homework to qualify for the Biometrician I register. I was a member of the Chief Fisheries Scientist's Office Statewide Shellfish Research Team for ten months, from January through October 1986.

I was given several assignments, including projects dealing with king crabs, a valuable commodity. King crab was Alaska's second most profitable seafood industry, just behind salmon. The record statewide harvest of 185 million pounds in 1980 declined to 16 million pounds by 1985. The economic impact of the fishery's collapse on Alaskan communities was extensive. Research of king crabs was vitally important to Alaska, and that is where my efforts were focused.

The foremost project was to re-evaluate historical king crab survey data collected in LCI to determine the precision of the relative abundance estimates, which had never been done. Reliable abundance estimates are crucial for managing fisheries that are based on a maximum sustainable yield strategy, as were king crabs. Faced with a declining king crab population of great economic import, the newly-installed biometricians in the Commercial Fisheries Division were eager to take a closer look at the surveys with me.

Past surveys assumed a random distribution of king crabs and an equal probability of finding crabs at each station. However, I discovered *aggregations* of king crabs. I don't know why the crabs were clustered in habitat patches. Environmental or social factors can influence their podding behavior and may intensify when the population is depressed. Regardless of the reasons, aggregated crabs present a problem because a random survey of an aggregated population leads to errors in the mean catch-per-pot. What to do? The answer: stratify.

Stratification is simply a way to group numbers into strata according to their similarity. When calculating the mean of each stratum, more precise mean values are obtained because there is less variability associated with them. The smaller the variability (or variance), the greater the precision, which improves the possibility of making correct inferences about the population.

I grouped stations with low numbers of king crabs into one stratum and high numbers of king crabs into another stratum, and calculated variance in the survey mean catch-per-pot per stratum. Using post-stratification to reduce variance produced estimates of mean catch that were less than the original estimates, indicating a significant overestimation of legal male abundance in some years.[46] For example, in the Kamishak District in 1975, the historical mean catch of legal males per pot was reported as 22, but using post-stratification, a more precise measure of the mean catch was 16.

Thus, due to a flaw in the survey design, overestimation of the legal male mean catch-per-pot resulted in greater than intended exploitation rates for king crabs in LCI during some years. This was a depressing discovery! We wrote up our analyses and conclusions, and met with the Region II regional supervisor and my former boss, Dick, to explain our concerns about high variances in relative abundance estimates using the current survey approach. We advised changes in the survey design. They replied, "We're not interested in variances." The regional supervisor then said, and I quote, "I'm not up for that science stuff."

I was amazed at their attitudes! Our advice was dismissed. For the next several years, I noticed no change in the LCI king crab survey design, nor was there any consideration of variance in the reporting of mean catch-per-pot. Given the personalities involved, I shouldn't have been surprised they ignored advice from a woman and a "number cruncher."

Just a few months into my new job, my boss, Dave, decided he wanted to work in the Sport Fish Division and transferred out of the Commercial Fisheries Division. The Statewide Shellfish Research Team was already disintegrating! I needed a new supervisor. There was talk of moving me to Juneau, but I found a biometrician in Region III, Linda, who was willing to assume my oversight, so I continued working for the Chief Fisheries Scientist's Office stationed in Anchorage.

My next project was to assist the manager of the Norton Sound king crab commercial fishery, Charlie, as an onboard observer. The role of the onboard observer is to ensure compliance with regulations, collect biological data, and radio updates of harvest estimates to the manager. The fishery opening was scheduled for August 1, so I flew to Nome at the end of July, a 90-minute flight from Anchorage. Located on the coast of the Seward Peninsula, Nome is accessible only by air. In ancient times, an Iñupiat settlement was built on the site. In 1898, gold was discovered in creeks and beach sands along the coast, which gave rise to the mining town of Nome.

Subsistence fishing for king crabs had long occurred in Norton Sound by its native residents. It was not until 1977 that the Board of Fisheries initiated a commercial fishery. Two fishing seasons opened: large vessels harvested crabs in summer, and locals fished through the ice in winter. The highest harvest achieved was 2.93 million pounds in 1979. The value of the fishery ranged from $229,000 to $1.9 million. The harvest strategy set an exploitation rate of 15 percent, half that normally inflicted on legal-sized male king crabs in Alaska, to protect the subsistence fishery.

Using survey and commercial catch information, the legal male population of king crabs before the start of the 1986 fishery was estimated to be 2.8 million pounds. At a 15 percent exploitation rate, the harvest quota was set at 427,000 pounds.[47]

Although the fishery opened on August 1, no vessels came! The fleet had decided to remain fishing for Tanner crab around St. Matthew Island. Finally, on August 12, two catcher/processors motored into Norton Sound, and two days later, a small catcher joined them, bringing the total to three vessels. Due to rough seas, Charlie could not take me out in his skiff to begin my observer duties until August 21, when the waters lay briefly quiet.

Norton Sound, an inlet of the Bering Sea, is cursed with shallow depth, on average just 45 feet deep. During open water in late summer and fall, low-pressure systems combined with southwest winds acting on the surface amplify waves into high lashings. As waves move from the deeper Bering Sea into the shallow shelf of Norton Sound, the energy released increases wave height, leading to increased roughness and frequency of breaking waves—not the best conditions for traveling in an open skiff.

Charlie wanted to install me on the small catcher. As we drew alongside, they lowered a rope ladder. I slung my little day pack across my shoulders and climbed aboard. I summed up the catcher as an old tub. The captain welcomed me with a big grin, flanked by fleshy jowls and rummy eyes. His First Mate was silent and kept to himself. The rest of that afternoon, they pulled pots, and I recorded their catch of king crabs, keeping a watchful eye to ensure they did not keep undersized males or females. I radioed the catch statistics to Charlie from the wheelhouse. Toward evening, the First Mate approached me with a big plate of food for dinner. The captain retained his grin. Shortly after eating, I became violently ill. I suspected at the time that I was poisoned. I threw up all night and was dry heaving the next day. With me incapacitated below deck, they fished the next day, and who knows what was tossed into their hold. On the evening of the second day, the waves became too rough to fish. As each wave slammed into the boat, its timbers shivered and sent me crashing into the walls, so symmetrical bruises formed on both sides of my body. With my energy nearly gone, I dragged myself up the steps to the wheelhouse, radioed Charlie, and said, "Get me off of this boat!"

Charlie arranged for me to transfer to the *All Alaskan,* a large catcher/processor from Seattle. The two captains planned to transfer me in the evening. They drew alongside each other, and the *All Alaskan* dropped a skiff over the side with their hydraulic winch. I was supposed to climb down the side of the old tub on the rope ladder, step into the skiff, and then the *All Alaskan* crew would winch the skiff up to their deck. I was mindful that their plan did not consider that an ocean swell could send one of the boats crashing into the other, smashing me in between the two vessels.

At the appointed time, I slipped on my little pack, swung a leg over the side of the old tub, and then momentarily hesitated. Going over the edge makes me catch my breath for a second, as my innate instinct to avoid danger fights it out with my brain that says, "It will be okay." I gripped the rope ladder, put one foot on a rung, swung my second leg over the railing, and found the next rung below. I lowered myself down to discover the rope ladder stopped short, about 8 feet above the skiff. I would have to push myself off the side of the old tub, let go, and drop in the middle of the skiff while I was swaying, and the skiff was bobbing in the swells. There was no going back. I pushed off, let go, and dropped. The skiff stayed afloat, and I didn't sprain my ankle in the fall. What a relief!

The skiff was winched up, and I was deposited onto the deck of the *All Alaskan*. I found an out-of-the-way corner. A young man glided by and handed me a small grape juice box. Mechanically, I took the cellophane wrapper off the plastic straw, punched it into the little tin foil hole, and sucked up a mouthful of juice. I had never tasted anything so delicious! This was the first nutrition I'd had in forty-eight hours. The sugar coursed through my blood and into my brain, which came alert. I saw an incredible ballet of perfect synchrony before me as the young men worked the crab pots. Every motion was fluid, never stopping, as pot after pot was raised, crammed full of king crabs that spilled onto the deck by the hundreds. I got out my clipboard, forms, pencil, and calipers and got to work—if they worked, I worked. We all ran on adrenalin that night. Occasionally, a crew member ducked into the galley for a quick bowl of Sugar Pops, then took his place again in the ballet. One young man told me that he expected his crew's share of the catch to be $50,000, and he planned to make a down payment on a house with the money. The oldest man on board was the captain, who was twenty-five years old, I guessed. He had the magic touch—he was on the crab! I was impressed by such incredible harvest power while mindful of the formidable task of sustaining crab populations subjected to this harvest power.

I radioed catch statistics every few hours to Charlie, who kept a close eye on the fishery. The next morning, the harvest quota had been attained, and he closed the fishery. The fishery had been a success, with an estimated ex-vessel value of $600,000. In the short time I was onboard, I measured 1,138 legal males, composed of 51 percent recruits and 49 percent post-recruits. The incidence of sub-legal and female crab catches was very low. In addition, I collected seventy-five tissue specimens for genetic stock identification purposes.

That afternoon, we watched the seas. If they were too rough for Charlie to retrieve me in his skiff, the *All Alaskan* would weigh anchor and leave Norton Sound. The Skipper threatened to drop me off at St. Matthew Island, an uninhabited rock in the Bering Sea. Thankfully, Charlie decided to forge through breaking waves to fetch me from the *All Alaskan.* As we headed toward Nome, I felt invigorated by the salt spray misting over me in the bright sun. After I stepped ashore, I found it funny that my sea legs had trouble adjusting to a firm surface. I stopped briefly at the office with Charlie to transfer data forms, then walked to the house of friends from Glennallen who had been temporarily transferred to Nome for wildlife research on the Seward Peninsula. They showed me to their guest room, and I crumpled onto the bed like a sack of dirty laundry. I slept twelve hours. The next day I boarded an Alaska Airlines flight back to Anchorage.

Seven months later, the *All Alaskan* shipwrecked on St. Paul Island in a gale. Thankfully, the crew was rescued. I figured I'd been lucky again to have dodged this catastrophe. My time in Norton Sound was my last maritime assignment with ADFG.

The collapse of the Statewide Shellfish Research Team was completed in 1986 when a recession caught Alaska by surprise. The recession was caused in large part by a worldwide surplus and drop in the price of oil. The price per barrel peaked at $35 in 1980 and fell to $10 in 1986. The plunge in oil revenue to the state caused large layoffs of workers. Simultaneously, a five-year program launched by ADFG in 1980 on the Susitna River (the "Su-Hydro" program) to collect data on fish, habitat, and water flow ended.

One day, when I was happily applying for the Biometrician I register, I was notified I was being "bumped" by a biologist who had been laid off from the Su-Hydro program. Bumping is a process agreed to in collective bargaining between the state of Alaska and the Alaska Public Employees Association whereby a layoff with seniority can take another person's job. The news was distressing! The person bumping me had no experience in shellfish or advanced statistical knowledge, so I had no idea what the Chief Fisheries Scientist would do with them.

A flurry of phone calls ensued. The Chief Fisheries Scientist said he would try to find money to keep me in a seasonal job but failed. The idea of moving me to Juneau was floated again. News of my dilemma must have traveled because I got a call from a colleague with the National Marine Fisheries Service in Kodiak, asking if I would consider applying for a position that was opening up with them. How kind of him to suggest this opportunity!

Not everyone was sympathetic to my plight. When my former boss, Dick, heard I was going to be bumped, he looked at me and said, "Damn!" and broke into a big smile. I knew then that he had been complicit in the ongoing campaign of discrimination in the Homer office.

The threat of job losses spread turmoil through ADFG. Some supervisors became inventive to protect their people. For example, the Sport Fish Division director transferred one employee out of the area and re-classified another, so Su-Hydro layoffs could not bump those biologists. As time passed, no rescue arrived, and my layoff date rapidly approached.

I received an evaluation from Dick that arrived several months late. After I had signed it, and without my knowledge, the Region II regional supervisor had scrawled on it in pen, "It is recommended that Ms. Merritt work especially hard on improving her working relationships with the crew of the vessel, *Pandalus*. Good working relationships are essential to the smooth and productive operation of a vessel where teamwork is vital to successful operations." Really?! I don't know what I must have done to improve my "teamwork." Should I have allowed the cook to jump on me, harass me, and hinder my research? Should I have allowed staff to threaten me, call me names, and demean me with porn plastered on the office walls? I had tried to fit in, be part of the team, and keep my head down. How did this help me in the end? Not much!

His ignorant statement was the last affront I was going to tolerate. I was angry! It was me against eight men in the Commercial Fisheries Division, coworkers and supervisors, who had created and condoned a discriminatory and offensive work environment. As long as I would lose my job anyway, I might as well tell my story. Perhaps my story would instigate change so other women at ADFG would not encounter harassment.

I had no idea what recourse was available to me, so I sought expert advice. I saw a management consultant and a labor lawyer, and with their help, I filed a grievance with the Equal Employment Opportunity Office in the State's Department of Administration. A few weeks later, I received a call that I had won my grievance. After reading my story, the Department of Administration removed the handwritten note of the regional supervisor from my evaluation and told ADFG to reinstate funding for my position.

I had raised the specter of gender discrimination in ADFG. Maybe some men would retaliate by refusing to hire me. However, I learned that no one would speak up for me except me. I resolved to persevere tactfully and professionally.

The director of the Commercial Fisheries Division acquiesced to the Equal Employment Opportunity Office ruling and allocated funding to retain my position in the Chief Fisheries Scientist's

Office. However, he placed a condition on it—a Biometrician II job would be cut. After mulling over this choice, the Chief Fisheries Scientist chose to retain funding for the Biometrician II, not me. I shouldn't have been surprised, yet I was. The bumping process would proceed.

My house in Glennallen needed repairs. I was sinking money into it for a new water heater, plumbing, paint, roof repair, etc. It was a money pit I could ill afford while facing a layoff, so I put the house up for sale.

On October 31, 1986, I was officially bumped from my position with the Chief Fisheries Scientist's Office. Working in concert with a few kind people in Region III of the Commercial Fisheries Division, I was able to slide into a position with no loss in service time. We contrived that I would bump a Fishery Biologist II in Region III, thereby taking his position as a Project Leader on the Yukon River, then supervisors would use seasonal money to keep the guy I had bumped on the payroll. In the end, three people were put through misery by being forced to change jobs, yet we all remained employed, so ADFG had not saved any money. The bumping process was an inefficient and stressful way to operate a workforce!

Yukon River Chinook Salmon

"You cannot escape the responsibility of tomorrow by evading it today." ~Abraham Lincoln

The mighty Yukon River is a 1,980-mile-long watercourse that originates in British Columbia, flows through the Yukon Territory, and continues westward through Alaska to the Bering Sea. Its remote and wild drainage area is 321,500 square miles. The river shifts tons of sediment to the ocean, and its unique mineral scent guides waves of salmon from rich feeding grounds in the Pacific Ocean back to their natal streams to spawn. The Yukon has one of the longest salmon migrations in the world, with Chinook, summer chum, fall chum, pink and coho salmon battling up tributaries of the Yukon River. As salmon do not eat on their spawning migration, those migrating the farthest have large fat reserves, especially Chinook salmon of Canadian origin, which are prized for their oily meat. The commercial harvest is one of the few sources of income available to Yukon River villagers, who also rely on subsistence fishing for salmon as food for humans and sled dogs. Native peoples of the river remark, "To us, salmon is life."

Commercial harvest of Yukon River Chinook salmon began in the early 1900s. The U.S. Department of Commerce began regulating the fishery around 1918. From 1918 to 1960, annual catches averaged 30,000 salmon. In 1961, ADFG took over, and the annual harvest surged, ranging from 63,838 to 158,018 fish. The increase was due to expanded effort—drift gillnet gear tripled, and fishing vessels doubled. The creation of the Commercial Fisheries Entry program in 1976 stabilized the amount of fishing gear through a permit system. In the early 1980s, the Yukon River Chinook salmon fishery was one of the largest in Alaska, with the average ex-vessel value of the harvest at $4 million.[48]

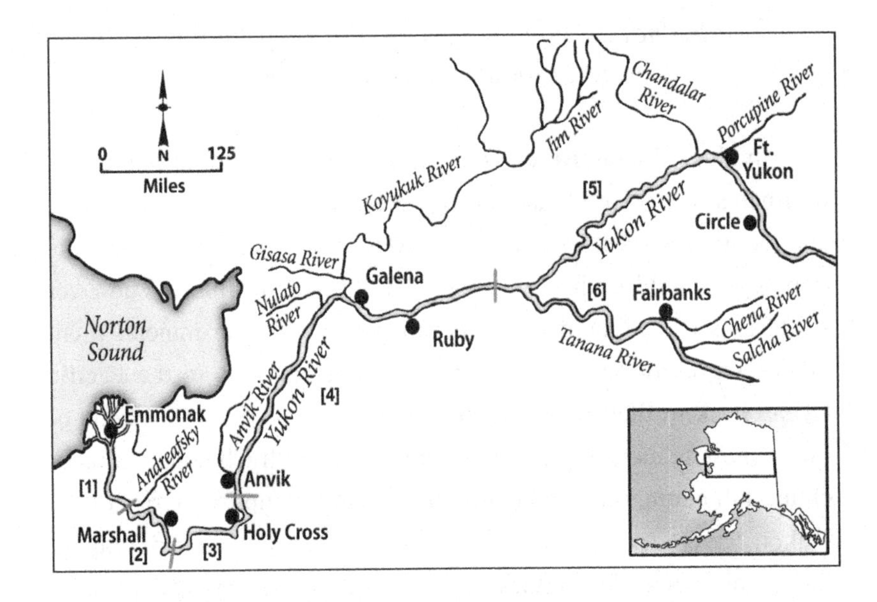

For management purposes, the Alaskan portion of the river was delineated into six regulatory districts. Most of the Chinook salmon commercial harvest occurred from early June to mid-July in Districts 1 and 2, where roughly 70 percent and 20 percent, respectively, of fish were taken in set and drift gillnets. The remaining 10 percent of the harvest was dispersed throughout the river.

Interception of the coveted Chinook salmon in various marine and freshwater fisheries was an early concern of biologists. During their ocean residence, they are targeted by foreign gillnetters and incidentally caught by trawlers operating in international waters of the North Pacific and the Bering Sea. The astounding catch of 704,000 Chinook salmon by the Japanese Mothership fishery on the high seas in 1980 intensified concern among fishery managers about the effects this magnitude of foreign interception has on stocks spawning and rearing in North American streams. Adult Chinook salmon of Yukon River origin that elude predation and capture at sea are targeted by Alaskan and Canadian subsistence, commercial, and sport fisheries along the river's entire length.

Conflicts arose as Alaskans took too many fish, leaving few fish for Canadians. Clearly, a legal agreement was needed between the U.S. and Canada to reduce the interception of Canadian-origin fish. The governments of the two countries negotiated conditions for a treaty. In March 1985, the U.S. and Canada ratified the Pacific Salmon Treaty, an agreement to cooperate in the management, research, and enhancement of salmon stocks of mutual concern. The treaty affirms commitments to prevent overharvest and ensure both countries receive the production of salmon originating in their waters. The Yukon River Salmon Agreement is specified in Chapter 8 of the Pacific Salmon Treaty. This agreement outlines steps to ensure the sustainability of Yukon River Chinook salmon stocks of Canadian origin.[49]

Under the Yukon River Salmon Agreement, Alaskan fishery managers are *obligated* to allow a target range of Chinook salmon to pass unscathed through Alaskan waters so they can reach Canada. The target range has to be achieved to meet Canadian escapement goals and harvest shares. The tricky part for ADFG in the mid-1980s was trying to figure out what proportion of Chinook salmon of Canadian origin were in the Yukon River at any given moment to limit their interception by Alaskan fishermen. At the same time, fishery managers wanted Alaskans to harvest the surplus salmon for their economic and cultural benefit.

To address the interception problem, fishery managers needed a tool to distinguish stocks for catch allocation and optimize yield from stock-specific production. In 1980, researchers selected a tool that looked promising—differences in the pattern of circuli on scales that reflect distinct growth histories of salmon from different portions of the Yukon River drainage. The tool was named scale pattern analysis.

My new job as Project Leader for the Yukon River Chinook salmon stock biology project was to estimate the interception of Canadian-origin fish by fishermen in the Alaskan portion of the Yukon River using scale pattern analysis. The information was needed to judge the compliance of the U.S. with the Yukon River Salmon Agreement of the Pacific Salmon Treaty. I felt a great weight of responsibility to be in charge of a project that had repercussions for our country's international diplomatic relations—not to mention the impacts to harvest allocation and local and international economies. I was anxious to perform well and meet everyone's expectations for accurate results. Thankfully, I was not on my own. Before authoring a report, I discussed decisions, explored interpretations, and debated conclusions with my colleagues. While I had confidence in the process we followed, with its technical reviews and statistical checks at each step, I was keenly aware that, as Project Leader, I was solely responsible for estimating Alaska's interception rate of Canadian salmon in the Yukon River. My professional reputation was on the line. No pressure.

As I stepped into my new job, I was handed a giant pile of scale samples. I already knew how to age salmon scales. Now I needed to learn how to measure the features used in scale pattern analysis. To measure scale features, I used digitization, which converts information into a computer-readable series of numbers. I spent most days that winter in a small dark room that housed the digitizing equipment, sitting at a large white tablet onto which I projected the image of a scale magnified 100 times. Measurements were made with a wand. I touched the tablet with the wand tip on scale features in a specific order and the tablet sent signals via a software program to the computer, which recorded the number of touches and distance between touches. While there were nine age classes of Chinook salmon, run-of-origin models were produced for just the predominant ages, 1.3 and 1.4. The repetitive nature of the work was monotonous, so scale digitizers devised ways to remain alert—we listened to music or books on tape.

I measured patterns in thousands of scale samples from salmon that had spawned in their home stream, building a data set that described typical growth histories. This data set formed the standards on which run-of-origin models were based. As it turned out, scale pattern analysis could not discern a salmon from a specific stream, but it could distinguish between salmon originating from three geographic areas: 1) streams of the Andreafsky Hills and Kaltag Mountains in the lower river, 2) Tanana River tributaries in the middle river, and 3) streams of the Pelly and Big Salmon Mountains in the upper river (Canadian portion).

For the next step in scale pattern analysis, I compared over 3,000 samples taken from fisheries in Alaska and Canada to the standards using statistical techniques and created models that classified the catch into the lower, middle, or upper river. The incidence of harvested fish classified to a geographic area was the mean interception rate for that group. In this manner, ADFG could follow, post-season, the mean interception rates of Canadian-origin salmon during fishing periods as the season progressed.

I explained the results to fishery managers so they could examine their management strategy. Of the total Alaskan harvest of 165,318 Chinook salmon in 1986, 26.6 percent was estimated to have originated from the lower, 5.5 percent from the middle, and 67.9 percent from the upper river areas.[50] Bad news! An exploitation rate at 67 percent or more was judged excessive, and if not reduced, would lead to a decline in abundance of Canadian-origin Chinook salmon stocks. The ADFG fishery managers had not anticipated such a high contribution of upper river fish to Alaska's harvest. They planned to adjust their management strategy next year to allow more Canadian-origin salmon to pass through Alaska and meet treaty obligations. Closing fishing periods early in the season would allow more upper river fish to escape to Canada.

For three days in March 1987, I attended meetings of the Yukon River Joint Technical Committee of the Yukon River Panel in Anchorage. The lead negotiators to the Yukon River Agreement had tasked the committee with hammering out mutually acceptable escapement targets, ways to improve the reliability of total return estimates, methods to identify depressed stocks, plans for stock rebuilding, and management approaches. The committee also reviewed and discussed the status of the stock identification program using scale pattern analysis. Besides myself, there were nine other ADFG biologists in attendance, along with representatives of the U.S. Fish and Wildlife Service, National Marine Fisheries Service, four biologists with the Canadian Department of Fisheries and Oceans, one member of the Canadian Territorial Government, and a representative from the U.S. Department of State.

During the meeting, the experts scrutinized my performance in digitizing and modeling scale patterns. Understandably, they had questions about the reliability of the scale pattern analysis program. With a coveted resource like Chinook salmon, both parties wanted their fair share. Fortunately, the committee members were a friendly bunch and worked well together.

When the 1987 field season arrived, I was ready to leave the digitizing room and explore new country. The plan was to sample salmon from the mouth of the Yukon River to its upper reaches in the Yukon Territory, Canada. In Alaska, I was assisted by technicians and biologists. In Canada, we were joined by Department of Fisheries and Oceans partners.

In early June, before the commercial fishery start date, I flew on Ryan Air from Anchorage to the small city of Emmonak, located 8 miles upriver from the vast mouth of the Yukon River, where it enters the Bering Sea. The Yukon River delta, at roughly 50,000 square miles in size, is one of the largest river deltas on earth. In 1987, Emmonak had a population of about 700 and boasted a hotel, a café, and a gravel airstrip. Home to the Yup'ik people, this flat marshland was once part of the Bering Land Bridge and had been occupied by humans for thousands of years. The Yup'ik live a subsistence lifestyle, hunting seals, moose, and caribou and fishing for salmon, whitefish, and blackfish. Their primary cash income is from the commercial salmon fishery. Emmonak is renowned for its variable weather, dependent on the Bering Sea. The locals say, "It can be sunny, cloudy, rainy, cold, or warm all in the same hour."

I stepped off the plane and walked across the gravel runway to a waiting ADFG truck. As we drove to the rented house that I would share with other ADFG employees for a month, I breathed in the complex smells of a river delta fishing community—rank estuarine vegetation, with a strong dab of rotting salmon, a whiff of sea breeze, and a hint of diesel oil. My June 7, 1987 journal entry read, "My suitcase and bike arrived two days after I did, but we all made it. Emmonak is either dusty or muddy, depending on the weather. No trees. The bushes don't have leaves yet." We rented the house from teachers who left in the summer and returned in the fall for the school year. I stayed in a little girl's cheerful room that had a sunshine quilt on the bed and toys strewn about. Having my bike was a big help in getting around town. My June 21, 1987, journal entry read, "It snowed last weekend. We have a few sunny days, but they are cool. The commercial fishery has been intense, and we work long hours, but I'm having fun and once again feeling good about my job. I work with considerate people—it makes all the difference."

Skiffs anchored at the Anpak processing plant in Emmonak, looking downriver to the Yukon River delta, 1987

Every morning, we went to the Anpak processing plant to sample the commercial catch from District 1 as it was offloaded from skiffs. Most fish were flash-frozen and packed onto barges headed for Seattle. To sample the District 2 catch, we traveled by air charter to the village of Marshall, a former gold rush town. District 3 was not sampled because few Chinook salmon were commercially harvested there. To access the catch in District 4, we air chartered to fish buyers. District 4 is a long stretch of river from Anvik to Tanana, encompassing thirteen villages of mostly Koyukon Athabascan residents. Districts 5 and 6 were accessed from Fairbanks. These districts encompass the villages of Fort Yukon and Circle, the traditional territory of the Gwich'in people of the Yukon Flats. The Han people are found near the village of Eagle.

To sample carcasses on their spawning grounds, we traveled to eleven streams in the Alaskan portion of the Yukon River drainage. I selected streams based on run size, geographic dispersal, and access. The time of sampling was gauged by the date of peak spawner die-off per stream. We obtained over 2,000 scale samples from spawning grounds.

The ADFG backup plan to scale pattern analysis was newly-evolving research of genotypic differences among salmon stocks. Thus, our assignment was to collect not only scale samples, but also tissue samples. In 1987, geneticists were uncertain which tissues provided the best distinction among groups of stocks, so we took tissue samples from the liver, heart, muscle, and fluid from the eyeball, a gooey and smelly job. Samples were placed into vials and stored in a dry ice container before shipment to the Anchorage genetics lab.

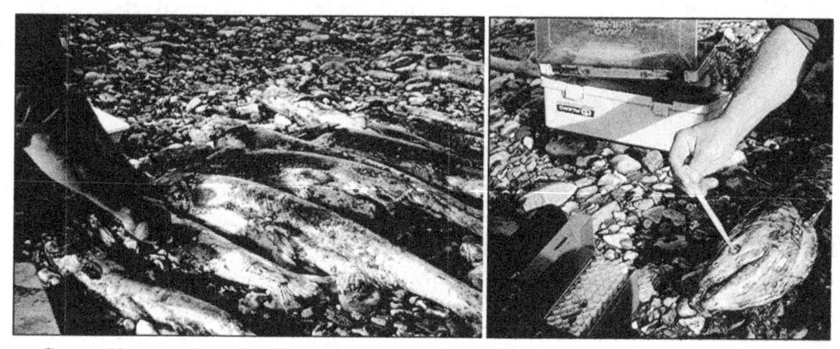

Sampling scales (left) and eye fluid (right) from Chinook salmon carcasses on the Chena River, 1987

Biologists with the Department of Fisheries and Oceans invited ADFG biologists to accompany them in sampling scales from Chinook salmon in the upper Yukon River. In the late summer of 1987, three of us flew to Whitehorse and met our Canadian counterparts. The Canadians were great fun and, in the spirit of camaraderie, presented us with T-shirts they had designed to commemorate our joint sampling venture. The design showed a Canadian beaver in a tug-of-war over a Chinook salmon with an American bald eagle, adding a light tone to an otherwise serious situation.

Logo on T-shirts gifted to us by Canadian colleagues

We spent a week traveling with our Canadian partners. The weather was gorgeous—bright blue sky, warm sun. I was impressed with the orderly gravel road system and camping facilities provided by the territorial government at the various waterways. The Canadian team used inflatable Zodiac rafts with floorboards powered by a 25-horsepower outboard motor to skip up shallow reaches in the Little Salmon, Nisutlin, Tatchun, Teslin, Morley, and Nordenskjold streams. We separated by pairs to cover more territory. In addition to sampling carcasses, we speared live spawned-out fish.

The Little Salmon River, Yukon Territory

I remember on one stream, my Canadian partner and I were hovering over a deep, clear pool in the raft, contemplating the dozen spawned-out salmon slowly ambling in a circle, waiting for the end of their life. I held tight to my spear, and to show off my spearing prowess, I struck with great force, hurtling the spear and myself into the pool. I surfaced from the biting cold water, sputtering to choking laughter from my partner. It took her some time to control herself enough to lend me a hand up and back into the raft. I sat sheepishly, dripping puddles, and joined in her laughter. My clothes dried out in the sun as the afternoon wore on. We met the others at a rendezvous point at twilight, where my spearing mishap was related with much gusto and renewed laughter. At our trip's completion, we had over 500 scale samples from Canadian spawning streams and judged our joint venture a resounding success with new friends.

Back in Anchorage, I settled into a second winter of digitizing scales and running models. In 1987, the commercial harvest was 202,125 Chinook salmon. Harvest of upper river fish decreased by 3 percent to an estimated 64.9 percent. The middle river fish contribution rose significantly compared to the previous year, to 18 percent. Lower river fish contributed 17.1 percent.[51]

With the advent of 1988, I was content. I was settled in Anchorage, working with good people and doing a job vital to an international salmon treaty. Jim and I got married. I finally sold my house in Glennallen. Although I was sad to lose the house I had worked so hard on, I was grateful for the years of memories with special people in Glennallen.

With little warning, my husband's employer transferred him to Fairbanks. I asked my boss if I could perform my duties stationed in Fairbanks, and he answered, "No." However, he kindly gave me six weeks of leave without pay to find a new job in Fairbanks.

I contacted a former Region III Commercial Fisheries Division biologist, Bill Arvey, who had transferred to the Sport Fish Division in Fairbanks. I asked him if he knew of any openings with ADFG in the Fairbanks office. As it happened, he was looking for someone to conduct stock assessment work on the Seward Peninsula, and he asked if I would be interested. Yes! I joined the Sport Fish Division in June 1988, where I remained for the rest of my ADFG career.

I was fortunate to have worked on the Yukon River during years when Chinook salmon were abundant. Since 2000, Chinook salmon stocks in the Yukon River have declined precipitously, endangering the economic and cultural welfare of thousands of people. The declining stocks are likely victims of an interplay of decimating forces such as a human-caused warming climate, interception, bycatch, competition, and overfishing. The Board of Fisheries classified them as a "stock of yield concern," defined as "a concern arising from a chronic inability, despite the use of specific management measures, to maintain expected yields, or harvestable surpluses, above a stock's escapement needs." Managers place strict limits on fishing if salmon numbers are too low to meet the threshold escapement judged necessary to provide a sustainable yield over time. The last commercial harvest of Chinook salmon in the Yukon River of any note occurred in 2010. The crisis has elevated emotions about who should have the right to harvest a now-scarce salmon resource.

Seward Peninsula Arctic Grayling

"I learned this at least by my experiment: that if one advances confidently in the direction of his dreams, and endeavors to live the life which he has imagined, he will meet with a success unexpected in common hours."
~Henry David Thoreau

In the spring of 1988, we bought a house in the hills north of Fairbanks, a short drive to the ADFG office. My new job would not begin until June 1, so I had time to explore the area. Fairbanks is the sub-arctic urban center of the sprawling North Star Borough, which at 7,400 square miles, is almost the size of New Jersey. In 1988, the city's population was about 30,000, with an additional 45,000 people scattered throughout the borough. Fairbanks sits at an elevation of 434 feet in a basin underlined with pockets of permafrost that can cause sections of roads and corners of ill-fated houses to slump with thawing ice. Bounded by the Alaska Range to the south and rolling hills laced with old mine tailings to the north, the basin is plagued in winter by temperature inversions that trap poor air quality. Extensive stands of spruce, interspersed with open forests of birch, aspen, cottonwood, and willow, blanket the landscape through which the silt-laden Tanana River flows.

Riches generated in the Klondike gold fields enticed prospectors to explore the mineral possibilities in Alaska, and a few gold prospectors made their way to the Tanana Valley. When Felix Pedro discovered gold in the area in 1902, the Tanana gold stampede launched a booming economy. The town came to be called "Fairbanks" when the newly-installed district judge, James Wickersham, suggested the site be named in honor of his mentor, Senator Charles Fairbanks.[52]

The ADFG office is located on the north edge of Fairbanks, adjacent to the 1,800-acre Creamer's Field Migratory Waterfowl Refuge, which used to be a dairy. Six divisions claimed space in the cramped single-story building. In addition to the Sport Fish Division, there was Habitat, Commercial Fisheries, Wildlife, Subsistence, and Administration. One wing served as laboratory space. Most people sat in small cubicles made from fabric-wrapped, moveable partitions, except for supervisors, who enjoyed windowed offices on the perimeter.

We used the dairy's old barn as storage space for our beach seines, gillnets, burbot traps, and other fish sampling gear. Pigeons roosted in the barn loft, so retrieving gear resulted in much wing flapping and pigeon whistling while we scurried to get in and out before being hit by bird poop.

On my first day in the office, I was put to work right away—there was no time to lose. I would head to the Seward Peninsula for the summer in two weeks. There were supplies to be purchased and shipped, sampling forms to assemble, and three technicians to hire. My job was to initiate life history research on Arctic grayling by surveying stocks in streams across the Seward Peninsula's 20,600 square miles, once part of the Bering Land Bridge that brought early man from Asia to northwestern Alaska during the Pleistocene Ice Age.

My objective was to discover basic growth characteristics. Key questions were, "At what rate do Arctic grayling grow?" And "Do growth rates differ among stocks?" This research aimed to identify streams that produce trophy-sized Arctic grayling and, using growth characteristics, develop regulations to ensure sustained yields for Arctic grayling fisheries.

The third week in June 1988, I flew to Nome. The familiar aroma of coastal air mixed with whiffs of treeless arctic tundra made me feel like I was home again. After greeting friends at the Nome ADFG office, I was given a vacant corner to pile gear.

I spread maps on a table and began to plan access routes, logistics, and sampling dates per location, with a target of sampling 600 Arctic grayling per stream. The waters we sampled from late June through mid-August were distributed across the Seward Peninsula as follows: the Nome, Snake, Sinuk, Pilgrim, Kuzitrin, Solomon, Eldorado, Niukluk, and Fish Rivers, and Boston Creek.

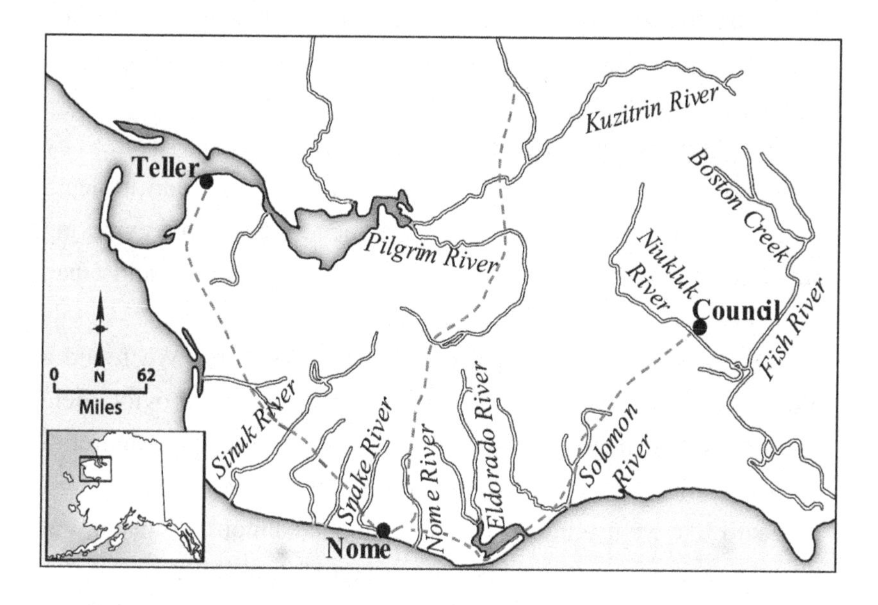

The Seward Peninsula is not connected to the rest of Alaska by a road system. However, three gravel roads maintained by the state from May through September originate in Nome and traverse the peninsula: the Kougarok Road (Nome-Taylor Highway) extends to the north, the Teller Road to the west, and the Council Road to the east. The roads, built in the early 1900s to transport gold to the port of Nome, provide 260 miles of access to the peninsula. Road accessibility to clear-water streams containing trophy-size Arctic grayling contributed to increasing sport fishing pressure near Nome.

In 1988, the Seward Peninsula supported the second-largest sport fishery in Region III. From 1983 to 1987, fishers averaged 18,764 angler days annually, casting for Dolly Varden, Arctic char, Arctic grayling, coho and Chinook salmon, and northern pike. A quarter of fish harvested (4,600) by anglers on the Seward Peninsula in 1987 were Arctic grayling, and many were trophy-size, defined as weighing at least 3 pounds and measuring 18 inches in length. The news spread when anglers discovered Arctic grayling on the Seward Peninsula that were 4 pounds and 22 inches in length. Increased publicity concerned managers because they lacked knowledge about Arctic grayling stocks.

Building on explorations of Seward Peninsula streams by biologists in the late 1970s, I was excited to dive into an in-depth study of Arctic grayling. Plus, I would spend the summer exploring new and wild country. My job was a dream come true, and I had earned it! I couldn't have been happier.

The most urgent issue I faced was transportation. We found a pickup truck to lease, and the troopers offered to provide field support by transporting personnel with a super cub. My watercraft of first choice was an inflatable Avon raft with a 25-horsepower outboard like we used for sampling Chinook salmon in Canada. For traveling across marine waters, we used an 18-foot skiff.

We would capture fish with two types of gear: a beach seine with small mesh measuring 100 feet long and 8 feet high to use in stream sections with slow to moderate current; and hook and line with barbless hooks to use in swift riffles and stream sections strewn with debris or boulders.

Varieties of sampling gear select for fish of different sizes. The beach seine captures smaller fish that rear in slow currents, while hook and line gear targets larger fish. Also in the sampling kit were maps, data forms, sequentially numbered Floy internal anchor tags, a tagging gun, a measuring board, tubs, gummed scale cards and tweezers for taking scales, and mosquito repellent. A shotgun was included for bear protection.

In the event we encountered anglers during our research, I included forms to record information on their harvest, fishing effort, gear, species and number of fish kept or released, demographics, and three questions regarding sport fish management:

1) How would you rate your fishing experience at this site?
2) Do you approve of recent regulations enacted by the Board of Fisheries to decrease the bag and possession limit to five Arctic grayling, allowing only one over 15 inches to be kept?
3) What is the smallest Arctic grayling you would keep?

I was struck by the difference in perspective between commercial fishing, which is primarily for profit, and sport fishing. Sport fish biologists consider nonmarket economic benefits, social issues, and opinions of anglers to better manage the relationship between humans and fish.

The three technicians I hired, Kirk, Kathy, and Tim, were great people to work with. Depending on the river system, we either worked as a team of four or split up to cover more country in groups of two. Sometimes my boss came with us on a sampling trip. A few rivers were close enough to town to allow us to run back to Nome for the night. Rivers farther afield required setting up campsites. The weather was mild, with sunshine the entire field season, except the last few days, when the parched peninsula received rain.

The first streams selected for sampling were close to Nome: the Nome, Snake, and Sinuk Rivers. The Nome River flows 40 miles from the Kigluaik Mountains to Norton Sound and is accessed from the Kougarok Road, which parallels much of its length. The river and its floodplain were heavily mined during the gold rush, and placer gold mining activity continues today. Clusters of summer and fish camp cabins dot the banks of the Nome River. We found deserted ruins and gravel piles above the river's banks, past evidence of the gold rush. I could see the effects of mining on the river. Fine sediments had settled in gravel spaces and compacted the substrate, reducing water flow and dissolved oxygen through the gravel, thus decreasing the survival of fish eggs buried there. Another effect of increased sediment loads from mining was an altered channel morphology.

We walked and rafted along the Nome River for eight days, making beach seine sets and casting our lines, hoping to catch Arctic grayling. With a total catch of fifty-four Arctic grayling, my goal to catch 600 fish per stream was not going to be met. Most Arctic grayling captured were adults; the youngest found were age-3. It was possible that young-of-the-year had little rearing habitat in the stream due to a combination of low water levels and alteration from mining and were swept into the sea. We theorized that while juvenile Arctic grayling are produced, their survival was low here.

I took a scale sample from each captured fish. Using a wooden board inlaid with a ruler, I measured the fish's length from its jaw to the fork in the caudal fin, called fork length. This method is a more accurate determination of length than total length, measured to the end of the caudal fin, because the caudal fin can fray. Fish measuring at least 150 millimeters fork length (5.9 inches) are deemed big enough to tag.

To tag a fish, I placed it in a tub of water and quickly inserted a T-anchor tag among the interneural rays of the dorsal fin with a tagging gun. I snipped the tip of the adipose fin with scissors to prevent double sampling of recaptured fish that might have shed their tag before releasing the fish into the river. We tagged fish to find out more about them. Recapture of a tagged fish can provide information on movement, age, and population levels.

I'm tagging an Arctic grayling in a tub

The Snake River is 36 miles long, with most of its watershed in low, rolling foothills succumbing to the coastal plain. The river was aptly named, with meandering bends looping back and forth, each with an inner sand bar and riffle of fast current at its end. The mouth of the Snake River ends in the Port of Nome.

Crew seining a river bend of the Snake River, July 1988

Along its upper reaches, the Snake River is accessed from a gravel road that ends at Jensen's Camp, an old mining remnant. We put in at the upper reaches and worked our way down to the bridge of Teller Road. A total of 265 Arctic grayling were captured here. I was happy to find good numbers of young-of-the-year fish (age 0) up to age 10, with most fish aged 5 to 8.

The Sinuk River flows 54 miles from the Kigluaik Mountains, entering the Bering Sea about 25 miles northwest of Nome. Sampling the Sinuk River was tricky. The only access points were the river mouth, entered from the sea, or at Teller Road Bridge. Unlike the Nome and Snake Rivers, there was no parallel gravel road, so fish were lightly exploited.[53] I decided to put in at the bridge. Motoring our raft upriver, we soon encountered large boulder fields that hampered further progress. Downriver, the stream braided into shallow channels presenting another challenge to travel. Low summer water conditions confined our sampling to a short stretch of the river. I was disappointed to capture few Arctic grayling here. It looked like fish had migrated to sections of the river that we could not access. I later learned that rafting the Sinuk River is most reliable in spring when the water level is high from snow melt.

Similar to the Sinuk River, we found few fish in the Solomon River. The Solomon River is a 22-mile stream accessed via Council Road. Prospectors heavily mined this river during the gold rush. Pick and shovels gave way to hydraulic nozzles that scoured gravels from the stream bed and benches on the hillside. In this tumbled landscape sits an old gold rush-era railroad engine with its disintegrating rolling stock and wooden track, a distinguishing feature of the Solomon River watershed.

The Pilgrim River, sampled from July 12-27, was my favorite trip. This river drains Salmon Lake and flows 55 miles into the

Kuzitrin River. Less damage from mining activity was evident here. Tim and I drove down Kougarok Road and put the raft in at Salmon Lake. We would take out at milepost 65, where the road crosses the river again, 32 miles downriver.

We had no advanced script on the river conditions we would encounter. This trip was pure exploration, and everything we experienced was new, wondrous, and unexpected. Our first surprise was finding high gravel terraces on either side of the river, towering above us as we floated by. I surmised the terraces were traces left by a glacier as it receded from the last Ice Age.

The second surprise was the sudden appearance of giant boulders downstream. The raft accelerated as it entered the rock-strewn stretch. We only had time to mutter a quick "Uh oh" before energetically paddling, trying to back-pivot the raft away from the first giant rock. As we spun around the first obstacle, the current slammed the raft into the next boulder. We were momentarily pinned sideways against the rock with water rushing on either side. We had milliseconds to react before the water would wash over the raft and push it under. Instinctively, we shifted our weight outward and leaned away from the boulder. Planting our paddles deep, we reached out to the force of the current and dug furiously, successfully twirling the raft off the rock and into an eddy on the rock's downstream side. We worked our way from eddy to eddy as we maneuvered through the remaining boulder patch. Whew! I was convinced that this stretch would not have been as strenuous if water levels had been higher that summer.

The Pilgrim River was calmer during the remainder of the trip. We took care to paddle a bit to the inside of river bends because water is pushed to the outside as it moves downstream. The trick is to stay out of the powerful push of the water and yet be in "controllable momentum" as you raft a river. We pulled up on the river bank at suitable spots to deploy the beach seine or cast a line, systematically sampling as we made our way downstream. At the end of each day, we scouted for a campsite with relatively level ground and firewood from surrounding cottonwood trees or scrubby willow. Most of the ninety-four fish captured in the Pilgrim River were big and among the oldest fish recorded, age-12. We found no fish younger than age-4.

We pulled out of the Pilgrim River and drove to the next stop along Kougarok Road, the Kuzitrin River. This long river originates at Kuzitrin Lake and flows 95 miles into Imuruk Basin. Along the road, I saw crumbled remains of sod houses used long ago by the Iñupiat for their fish camps. Another interesting feature along the road is the steel truss Kuzitrin Bridge. Originally built in 1917 for use in Fairbanks, the bridge was disassembled during that city's bridge renovation project and gifted for use to Nome. The pieces were barged down the Chena, Tanana, and Yukon Rivers and up the coast of the Bering Sea to Nome. From there, workers hauled the pieces to the banks of the Kuzitrin River and reassembled the bridge in 1958. After only two days spent deploying the beach seine and casting line, 195 Arctic grayling were captured, ranging from young-of-the-year to age-10 fish—a good haul!

It was now August, and I could smell fall in the air. The tundra was turning red, berries were ripe, and soon animals would be migrating. During August 5–6, I sampled the Eldorado River with a biologist from Fairbanks, Fred, who, in later years, took over Arctic grayling research on the Seward Peninsula. The Eldorado River is 14 miles east of Nome and is accessed via Safety Sound. The 30-mile-long river supports a large escapement of chum salmon which supplies a nutritious source of eggs for voracious Arctic grayling. From Safety Sound, we navigated the skiff into a narrow inlet and kept to the right to enter the river's mouth. The lower section snaked around small ponds framed with grasses and sedges before entering a broad gravel-filled valley. Due to the size of the skiff, necessary for traveling across the sea, we were limited in upriver travel. It was tough for us to fool the Arctic grayling into hitting at our lures when they could eat salmon eggs. Despite the challenging fishing, we captured fifty-four fish, most of which were age-4 and age-5.

The Niukluk River

The last waters sampled were the Fish River and its tributaries, the Niukluk River, and Boston Creek. This river system begins in the Bendeleben Mountains and flows to Golovin Bay in Norton Sound. While my boss and another biologist flew in to sample Boston Creek, my crew and I traveled the Council Road to survey the Niukluk and Fish Rivers for two weeks. After a summer of clear skies, clouds moved in, bringing rain to our last sampling trip. The Nome-Council Road is 73 miles long, and ends on the south bank of the Niukluk River. The old mining town of Council sits amidst shrubby tundra on the north bank of this river. We put in our raft where the road ended and worked our way downstream to where the Niukluk enters the Fish River. We captured 284 fish in this section of the Niukluk River. We found only a few fish age-3 and younger; most were age-5 and age-6.

The Fish River is much broader and deeper than the smaller streams we surveyed. In addition to Arctic grayling, Dolly Varden, whitefish, and salmon, fishers also caught burbot and northern pike here. Fifteen miles downstream from the confluence sits the village of White Mountain on the site of an ancient Iñupiat fish camp. With no road access to the village, residents either boat to Council, then drive the road to Nome, or catch a flight from the small airstrip adjacent to the village. I motored over to the east bank to visit a sport fishing lodge operating in the village and speak to residents about fishing. We caught sixty-five Arctic grayling in the Fish River. Similar to findings in the Niukluk River, we found only a few younger fish; the majority were age-5 and age-6.

We broke camp in the rain on the last day in the field. Packing sodden gear in the raft, we motored upriver to where our truck was parked on the south bank of the Niukluk River and drove to Nome. In six weeks, we sampled 990 Arctic grayling in ten streams. As I reflected on the summer's work, an old saying came to mind, "Choose a job you love, and you will never have to work a day in your life."

I had encountered thirty-two anglers, most of whom fished the Nome and Niukluk Rivers, but there were a few at other streams. The anglers were primarily local residents who fished for pink salmon, followed by Dolly Varden and Arctic grayling. Their opinion of the quality of fishing relative to previous years was mixed: 47 percent claimed it was the same, while 33 percent and 20 percent declared it worse or better, respectively. The majority thought more restrictive regulations for Arctic grayling were a good idea. Nearly half the anglers interviewed said they preferred to keep Arctic grayling greater than 18 inches in total length.

About a third of the way through the summer, I happily discovered I was pregnant and looked forward to starting a family with my husband. The pregnancy did not interfere with my fieldwork. In the morning, I simply stepped behind a tree, threw up, and began that day's river survey with my morning sickness out of the way. I didn't tell anyone in ADFG as I figured a pregnant biologist was new ground for the men and they might freak out by mistakenly viewing pregnancy as a debilitating illness. When I was back in Fairbanks and told my boss I was pregnant, Bill asked,

"Why didn't you tell me?"

"Because I thought you might take me out of the field. Would you have?"

"Yes!"

"Well, then, that's why I didn't tell you!"

I hoped I had proved my point—pregnant biologists can perform fieldwork and get the job done. I knew I was setting an example, and future female biologists might be judged by my performance. I regret not confiding in one of the technicians, Tim, about my pregnancy because over the course of the summer he became increasingly alarmed about my morning vomiting routine. When he finally mustered the courage at the end of the summer to suggest I see a doctor, I laughed and reassured him I was fine, just pregnant. I was touched that he was concerned about my well-being.

Back in the Fairbanks ADFG office, I aged scale samples and analyzed data collected over the summer. Growth slows after reaching maturity in long-lived fish, such as Arctic grayling, so outer annuli on scales become indistinguishable. In addition, the edges of older scales fray. This means that annuli after age 12 are difficult to discern. In my research on aging Arctic grayling conducted in later years, I discovered that scales underestimate the age of older fish, so we likely found fish on the Seward Peninsula older than age 12.[54]

Young age classes were underrepresented for most rivers surveyed, possibly because: 1) small fish inhabited side channels that we didn't sample; 2) reduced winter habitat may force small fish into areas inhabited by large fish, increasing their risk of being eaten; or 3) recruitment is limited by flood events during critical times of fry emergence and rearing. Streams on the Seward Peninsula likely experience high spring water events that wash young fish into the sea. Strong and weak cohorts would be expected in a population if periodic river discharge was limiting recruitment.

That is what I found! Many stocks were composed of just one or two age classes. Assuming river coverage and capture methods were sufficient to sample all age classes, it appeared that only a few strong cohorts supported fisheries for Arctic grayling in these rivers. This finding was not unusual. Arctic grayling in the Tanana Valley were also found to have one or two strong age classes that supported most of the fishing pressure. Reliance on periodic strong year classes makes these stocks more vulnerable to overharvest.

I had enough samples to calculate growth curves and was able to affirm that growth characteristics vary among Arctic grayling stocks on the Seward Peninsula.[55] Arctic grayling from the Nome and Snake Rivers grew faster between ages 4 and 10 than fish from the Fish and Niukluk Rivers. Why did fish growth rates vary in different rivers? Were the observed differences related to genetic or environmental factors? Often, research to answer one set of questions leads the investigator to ask other questions.

After finishing my report on Arctic grayling research on the Seward Peninsula, I went on maternity leave. I planned to take three months off. I had six weeks of sick leave saved, and the remainder of the time, I would be on leave without pay. However, my leave was unexpectedly cut short.

Research Supervisor

"Try to understand who you are and what you want to do. Don't be afraid to go down that road and do whatever it takes and work as hard as you have to work to achieve that." ~Sally Ride

I was at home on maternity leave when I got a call from the regional supervisor, Dr. John H. Clark, asking if he and the research supervisor, Rocky, could come to talk with me. Mystified, I said they were welcome to visit. I ushered my guests into the dining room, and we sat at the table, my tiny three-week-old daughter in my lap, asleep. John told me about the environmental catastrophe from the grounding of the tanker *Exxon Valdez,* which had strayed from shipping lanes to avoid icebergs and hit Bligh Reef on March 24, 1989, spilling 10.8 million gallons of crude oil into Prince William Sound. The oil affected 1,300 miles of coastline and resulted in the immediate deaths of over 100,000 seabirds, thousands of marine mammals, and untold numbers of fish and crustaceans.[56]

The rupture of the tanker's hull occurred just after midnight. ADFG supervisors were called at home in the dead of night to inform them of the accident. The Commissioner of ADFG had summoned "all hands on deck," so John and Rocky were traveling to the affected area. Biologists from all over Alaska were being routed to Cordova, which became the headquarters for response to fisheries concerns. Response to the oil spill consumed the life of biologists for months; for some, the commitment extended into years. Immediate tasks were identifying critical habitats and resources being affected and directing initial cleanup efforts toward sensitive areas.

In urgent tones, John requested that I come back to work. With supervisory staff in Cordova, my job would be to oversee the completion of Federal Aid in Sport Fish Restoration studies to meet federal reporting deadlines. Seventy-five percent of the Sport Fish Division budget came from federal funds derived from taxes on fishing gear and boat fuel, designed to increase sport fishing and boating opportunities through research, management, and restoration projects. This funding would be jeopardized if the region didn't meet reporting deadlines.

My first response was that I couldn't—I had a baby to take care of! John asked me to think about it. If I agreed to oversee the research program, I would be designated acting research supervisor at the rank of Fishery Biologist IV. Rocky was considering a promotion and transfer to Juneau, so I would eventually be in contention for his permanent position. A sticking point to hiring was that I needed credit working as a Fishery Biologist III to get on the Fishery Biologist IV register. John asked, "Have you ever worked in the capacity of a Fishery Biologist III?"

"Well, yes," I replied, "I did the same duties as a Fishery Biologist III in Homer but was only assigned a Fishery Biologist II rank."

I told him the whole story. John suggested I petition the Division of Administration for credit as having worked as a Fishery Biologist III in Homer, so I did. Division of Administration staff reviewed my petition, acknowledged I had worked in the capacity of a Fishery Biologist III, and gave me credit for the time. However, I was not awarded the pay difference that accompanied the credit. I became another statistic of a woman working the same job as a man but getting paid less. As long as I was given credit for the job I had done, I was ready to put Homer behind me.

My name was accepted to the Fishery Biologist IV register. Following application and interviews, I was hired, becoming the first woman to hold the rank of Fishery Biologist IV in ADFG's regulatory divisions (Sport Fish, Commercial Fisheries, and Game). We agreed that instead of taking three months off for maternity leave, I would cut my leave short and return after two months. I could work through lunch and get home to hold my baby from 4 p.m. to 10 p.m. Fortunately, I found a wonderful babysitter to come to our home while I was at work.

John had a persuasive manner, and holding a doctorate, he was also a valuable advisor. John convinced me that my ability to perform my duties and my credibility as a research supervisor would be enhanced if I completed a doctorate degree in Fisheries—while I was doing a full-time demanding job as well as functioning as a wife and mother. John allowed a small portion of my time from work to attend classes, but I paid for them. I contacted the Fisheries Leader, Jim, at the Alaska Cooperative Fish Research Unit, about the process of applying for a Ph.D. degree in Fisheries. Jim helped me assemble a diverse and insightful committee of five members who outlined my curriculum and research. Thus, I became exceedingly busy and adept at precise time management for the next five years.

When I walked back into the Fairbanks office in May 1989, the area encompassed by Region III Sport Fish was about 409,600 square miles in size or 70 percent of Alaska's land mass. Eight years later, the region would expand to 526,000 square miles. The region included the Arctic and Yukon and Kuskokwim River drainages, so it was also called the AYK Region. Thousands of lakes and miles of streams supported many recreational fisheries for freshwater and anadromous species. These species included Arctic char, Dolly Varden, Arctic grayling, northern pike, burbot, lake trout, sheefish, all five species of salmon, and several whitefish species. The majority of angler effort and harvest occurred near population centers in the Tanana River Valley. Although Region III was vast in size, it was sparsely populated. During the early to mid-1990s, Region III accounted for only 7 to 10 percent of statewide angler effort and 4 to 6 percent of statewide harvest of sport fish, as estimated by the annual postal Statewide Harvest Survey.

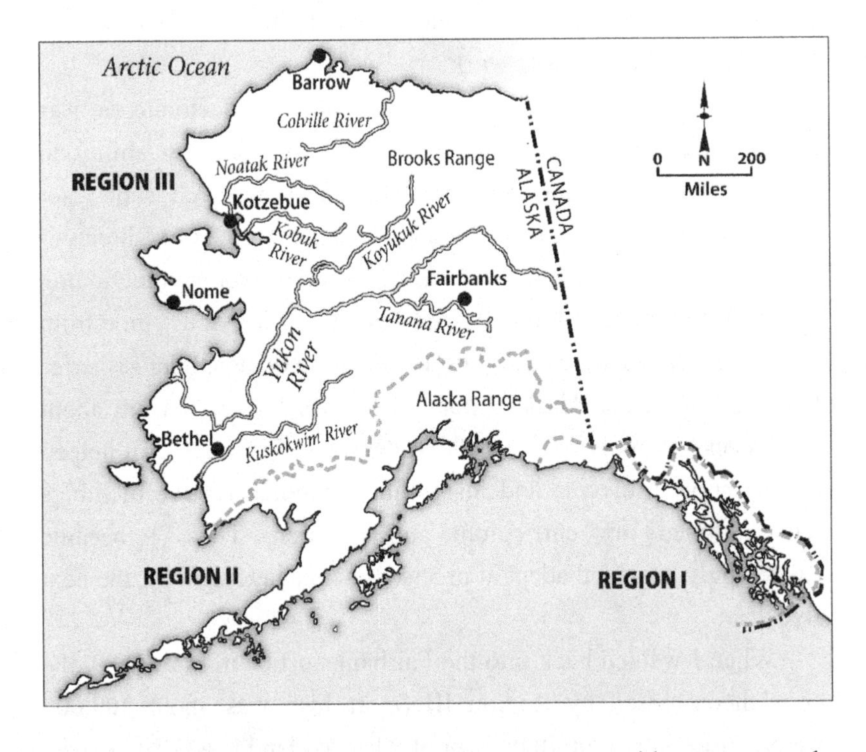

My job as the research supervisor was to guide ten research biologists, and their field technicians, in planning research projects in consultation with biometricians. Area managers also often engaged in research, and although not under my direct line of command, I supervised their research activities. Facilitating project implementation, data analysis, and reporting consumed a large part of my time, along with monitoring the budget.

The seasonal cycle of the job went like this: in spring, there were operational planning meetings, end-of-fiscal-year budget adjustments, and editing of belated reports hastily turned in by biologists before another field season began; in late spring, technicians were hired and needed equipment purchases were made; in early summer, project leaders and their teams began fieldwork; at the end of summer to early fall, field projects finished, technicians were put on leave, and biologists cleaned, repaired, and stored equipment. Winter was the time for data analysis, report writing, and editing before early spring, when the cycle began again.

In my first year, I got an idea of how intense work at the Fishery Biologist IV level was. In 1989, I co-authored a management report, edited twenty-one research reports with the help of my Publication Technician, Kerri, lent a hand on field projects, and supervised the creel survey.

The creel survey project concerned seven major sport fisheries in the Tanana River drainage.[57] While the Statewide Harvest Survey was a cost-effective way to obtain general statistics on harvest and participation, on-site creel surveys allowed the timely collection of detailed information to regulate sport fisheries better.

Creel surveys monitor angler effort and harvest, gather biological data on fish harvested, describe angler demographics, assess the consequences of regulations, evaluate the effectiveness of stocking programs, and obtain angler opinions. To conduct a creel survey, fishery technicians visit a fishing site during assigned periods throughout the season. They either stay at a popular access point or patrol the fishery by foot or boat.

Few fishery technicians were available because of the need for extra hands in Prince William Sound to deal with the oil spill, so I joined the survey crew. Some fishing sites close to town transformed into rowdy party areas at night. To increase safety during surveys of such sites, I assigned two technicians to patrol and handed out radios so they could communicate with each other and me. Cell phones were not yet in wide use.

A fun fishery to survey was the Chatanika River whitefish spear fishery. The Chatanika River supports fall spawning runs of primarily humpback whitefish, least cisco, and to a lesser extent, round whitefish. Historically, anglers pursued whitefish with rod and reel, but the fish were difficult to catch this way. Anglers began to seek other ways to harvest whitefish. The result was the establishment of a spearfishing season for whitefish in the Tanana River drainage in 1969.

To spear whitefish, fishermen wade into the icy river at night, when fish are likelier to be on the move. Clutching a 10-foot spear with barbed tines in one hand and a lantern in the other, fishermen search for fish flitting past their feet. For many, the activity arouses a primordial feeling, such as primitive man stalking prey, albeit in neoprene waders. Being fall, the temperature was cold, and aufeis was often found at the river's edge. Families set up campsites and got a roaring fire going at which the spearfishermen could periodically warm up before plunging back into the river.

The spear fishery developed slowly but, over time, became popular. In response to the rapidly increasing whitefish harvest in the early 1980s, several methods to assess whitefish in the upper Chatanika River were explored, such as sonar, tower counts, and mark-recapture experiments. Mark-recapture proved the most cost-effective and precise method to estimate composition and abundance.

To do a mark-recapture experiment, fish are captured, marked, and released back into their habitat. After some time to allow marked fish to mix with unmarked fish, another portion of the population is captured and examined for marks. Since the number of marked fish in the second sample (recapture event) should be proportional to the number of marked fish in the entire population, an estimate of abundance is obtained by dividing the number of marked fish by the proportion of marks found in the recapture event. The success of mark-recapture depends on meeting several conditions: marking doesn't affect catchability, marks are not lost between sampling events, there's no death or recruitment between sampling events, and marked fish mix thoroughly with unmarked fish. Most of the time, if one of these conditions is not met, the data can be corrected to produce an unbiased estimate.

The Chatanika River mark-recapture of whitefish was the last field project for the season, and many of us in the office looked forward to lending a hand to the project leader, Saree, in sampling fish.[58] One last foray into the wilds before we were stuck behind our desks for the winter! The Chatanika River is beautiful in the fall, with golden birch leaves dripping over the gently rippling water. As I came to learn, the river has some sharp bends in shallow water, so you have to expertly steer the jet boat to stay off the bank.

From population assessment research, biologists thought a harvest of 10,000 whitefish would be a sustainable yield. To achieve the targeted harvest level, managers instituted a fifteen whitefish daily bag and possession limit. While managers hoped the restrictions would constrain harvest, there are no guarantees when second-guessing strategic human predators.

The fishery occurred in three areas, so for the creel survey, we had roving technicians greet people while they fished, and one technician conducted exit interviews as people completed their fishing trip. The creel survey ran from September 9 to October 16, and each survey period lasted six hours, between 8 p.m. and 2 a.m., when the majority of spearfishing occurred. On my shifts, I recall that the black sky glowed with infinite stars, sometimes obscured by lime green northern lights shivering across the sky. I stayed warm, bundled up in wool and down, with hand warmers. I enjoyed the festive atmosphere as local families enjoyed an evening out on the town, Fairbanks-style.

In 1989, spearfishermen harvested just over 16,000 whitefish. It looked like the fifteen-fish bag limit was not controlling the harvest to the expected 10,000 fish per year, so further restrictions to the fishery were implemented.

I made it a point to get in the field periodically to see research projects unfold and address any issues. Plus, it was good to get away from my desk and enjoy Alaska's outdoors. In my first year as the research supervisor, I was keen to see electrofishing projects and become more familiar with them. Electrofishing is used to capture fish when other methods are ineffective, such as in fast-flowing rivers or marshes.

Electrofishing uses pulsed direct current, powered by a gas generator or battery, to create an electric field between a cathode and anode submerged in water. When fish encounter the electricity, they become momentarily stunned. As the electrical field dissipates, they recover to dash off, so electrofishing requires one person to operate the anode while dip netters catch fish. Stunned fish are quickly scooped into tubs of fresh water, sampled, and released. The effects of electricity on fish depend on such factors as the strength of the electric field, how close fish swim to the anode, and fish size. The biologist seeks the minimum electric field (threshold) needed to achieve capture.[59]

A crew working with electrofishing equipment spent time in training and were outfitted with electrician's gloves and neoprene or rubber waders to guard against electrical shock. Verbal signals were given to let everyone know the anode was live. Because high voltages in water pose a serious safety risk, at least one crew member was certified in cardiopulmonary resuscitation.

Biologists used both boat-mounted and backpack electrofishers. In boat operations, there was a crew of three: two dippers who stood on the bow, enclosed by a railing, and a driver, who also operated the electrofisher. Anodes were either steel cables or spheres arranged forward of the bow. The unpainted bottom of the aluminum boat was the cathode. The driver adjusted voltage, duty cycle, and pulse rate to the water conditions, species of fish, and their reactions to the electric field to find the threshold. For wide rivers, two electrofishing boats traveled downstream, one along each bank, to collect as many fish as possible in a twenty-minute interval.

I remember crewing as a dipper on Interior rivers. Keeping balance on your feet while standing on the bow of a moving boat, ducking away from overhanging branches, and leaning over the railing to scoop up a stunned fish before it skitters out of arm's reach is harder than it sounds! We soon got into the rhythm of swinging the pole aboard the boat as we quickly twisted to drop a netted fish into the holding tank without accidentally hitting the other dipper with the end of the pole. At regular intervals, the generator was turned off and the anchor set so we could process and release fish that had accumulated in the holding tank. The purpose of these studies was to use the mark-recapture method to estimate population abundance, age and size compositions, recruitment, and survival for the assessment of regulatory measures.

Sampling small streams called for a wading operation, where two people with backpack electrofishers, accompanied by four dippers, wade abreast to cover the stream's width as they progress down-current. I joined crews in capturing Arctic grayling[60] and rainbow trout[61] in Piledriver Slough, a small stream close to Fairbanks. This stream was inadvertently created in 1976 by blocking the Tanana River's flow, thereby exposing a spring. Various fish colonized the stream, including Arctic grayling. Sport fishing in Piledriver Slough became very popular with local anglers, especially after a rainbow trout stocking program began in 1987. We needed to assess the Arctic grayling population and evaluate the stocking program's success through mark-recapture studies.

I carried one of the backpack electrofishers and held the anode, attached to the end of a 6-foot-long pole, in one hand as the cathode, a 9-foot-long braided steel cable, trailed behind me. On the pole was a "deadman's switch" to cut off the electrical current, set at 60 Hz pulse DC ranging from 200 to 250 volts; the amperage varied from 1.5 to 2.0 Amperes. I watched my feet carefully as I waded downstream. I did not want to step into a deep pool, slip on mossy rocks or become entangled in tree roots with that heavy unit on my back. I moved the pole in an arc. Dippers stayed close to the submerged anode, ready to scoop up a fish. Captured fish were placed in tubs we lugged along for sampling, then released into areas not under further electrical stimulation.

There are two problems with using electrofishing as a capture method. One problem common to all capture methods is size selectivity. A fish's vulnerability to electroshock increases as its size increases. In contrast, smaller fish require higher threshold voltages. This problem might result in larger fish being over-represented in the sample. As long as the biologist is aware of the problem, statistical procedures can be used to correct the bias.

Electrofishing rarely harms the fish when performed correctly, but there is potential damage to the fish from excessive electroshock. In fieldwork, we noticed that on some occasions, fish became discolored following electroshock, suggesting internal damage. To learn more about the effects of electrofishing on sport fish in Region III, we turned to a wonderful program conceived by the Sport Fish Division in partnership with the Alaska Cooperative Fish Research Unit. The Division funded two to three qualified Sport Fish Division employees annually to participate in graduate research related to ADFG priorities. Often, the graduates of this program developed into high-ranking biologists. Two Region III employees, Tom and Don, were selected to research electrofishing. Their examination of spinal injury, internal bleeding, survival, and growth of fish exposed to varying levels of electric current under various conditions helped to direct ADFG's electrofishing program in reducing fish injury while improving capture efficiency.[62, 63]

During 1990, work grew more intense, with some biologists responsible for five field projects. With the help of my new Publication Technician, Sara, I edited thirty-four reports, researched fish aging techniques, and assisted with field projects. The workload was stressful for everyone.

In 1993, John transferred to the Chief Fisheries Scientist position in the Commercial Fisheries Division, and Region III Sport Fish acquired a new regional supervisor who got off on the wrong foot with a few of us. I was relegated to an "office" that measured about 6 feet by 8 feet. It was a small space. I had to squeeze my butt into the wall to get behind my desk, and there was only room for one person to sit in a chair across from me with their knees drawn up. The space was so claustrophobic that biologists I supervised preferred to meet with me in the hallway.

One day, I learned my boss had re-directed a crew of research technicians without my knowledge, messing up the project biologist's work for the day. In mark-recapture projects, days elapsed can be critical. I stomped into his office, slammed the door shut, told him he would no longer screw with the research staff, and then told him why—carefully planned research can be compromised when he interferes without considering the consequences.

Thereafter, when he wanted an extra hand, he asked first about research staff availability. As for my office problem, I worked directly with the Division of Administration to get a larger space.

As my daughter, Laura, started kindergarten, I received permission to shift my work day by 1 hour so I could take her to school in the morning and get to work at 9 a.m. It meant the world to me to be a mom in the morning and see my child safely off to school. Integrating flexibility needed for parenting was a novel concept at the time, and only some of the guys were comfortable with it. There was a suspicion that flexibility of the work day for childcare meant getting privileges no one else had and that a "lax" schedule would lead to less work. Frankly, motherhood was not viewed as masculine—a valued trait in the workplace. One day, when I asked a question, a man snapped at me, "Well, you'd know if you came to work at 8 o'clock like the rest of us!"

I ignored the barb and calmly juxtaposed my work and personal lives. On days when the babysitter was unable to pick my daughter up from school, I fetched her and installed her at a little desk I had sequestered in the corner of my office, where she had ample books and toys to keep her busy while I continued my work day. I had triple back up child care to ensure care was covered when I was in the field.

I tested the staff's tolerance by taking two and a half hours of leave a week (when work permitted) during my daughter's first winter of school to volunteer in her kindergarten class—it was so fun! I learned a lot about the vagaries of early childhood development. As the staff realized that these minor adjustments in my schedule did not hamper the region's productivity, I think they resigned themselves to the situation, at least outwardly, and we continued in the business of sport fish stock assessment and management.

Stock Assessment in Region III

"After much experience in the field, we would be the first to agree that it is indeed impossible to capture fully the rich behaviors of ecosystems in mathematical models, particularly when we try to include humans as dynamic predators in the calculations. But... it is important to keep trying." ~Carl Walters and Steven Martell

Changes in the composition and abundance of a fish stock are a result of dynamic natural and human factors, which in turn can impact other creatures in the aquatic food web. In the face of this complexity, it seems a bit foolish to think we can accurately assess a single species at a point in time and then predict what will happen to that fish stock when people selectively harvest it. Nonetheless, managers *must* respond to human harvest by making policy choices. The role of stock assessment is to piece together biological information to help managers make correct choices for preferred outcomes. Examples of stock assessment studies in Region III and their outcomes appear below.

Northern pike

Anglers in the 1960s considered northern pike a pest. Perception of the fish changed over time, and northern pike became a popular game fish. The estimated sport harvest of northern pike in Region III in 1989 was 17,100 fish, with most taken in the Tanana River Valley. Several biologists assessed northern pike populations in such waters as Minto Flats, George Lake, and the Yukon River.

Region III Sport Fish owned a Cessna 185 on floats, and we were fortunate to have one of the northern pike biologists serve as a pilot. When not in conflict with his research duties and on a day with good weather, I asked Gary to fly me to one of the many research projects in the field. One visit, in particular, comes to mind. In June 1992, we flew to George Lake to check on the progress of northern pike studies there.

George Lake is 4,500 acres in size with shallow beds of aquatic vegetation that is the perfect spawning and rearing habitat for northern pike. The recent estimated annual harvest was 1,722 northern pike. To ensure the harvest was sustainable, biologists estimated the abundance and composition of the northern pike population 12 inches in length and longer. Twelve inches is the length at which pike are fully recruited (susceptible) to capture.

It was a beautiful sunny day for a flight, and the plane landed on the lake perfectly, softly skidding to the bank where it would be tied for a few days. We hopped out and headed to the crew cabin, rented for the study's duration. The area manager for the Upper Tanana drainage was the project leader this year; he was making pancakes for breakfast. We had arrived at an opportune time. After breakfast, we started sampling the lake's perimeter with a big bag seine, measuring 216 feet long and 10 feet deep. The net was cast from a boat in shallow feeding areas, close to the lake's edge, and retrieved by hand to shore. Fish were sampled, tagged, and released.

In the morning I helped pull the net to shore and sample fish. Northern pike have sharp teeth in the jaws, roof of the mouth, tongue, and gill rakers, so we were careful where we placed our hands when reaching into the net to secure a squirming fighter. Toward mid-day, I switched places to help the boat crew. After the first set, the project leader announced a "tradition" of George Lake: biologists new to study there were "christened" by being tossed into the lake. I narrowed my eyes and looked at him sternly, saying that sounded like a stupid tradition, whereupon he came toward me determinedly. The crew looked on in askance, unsure if it was a sane idea to toss your supervisor into the lake, but they stood by mutely while Fronty hefted me overboard. We were in shallow water, and I landed on my feet so I got only partially soaked. For a moment, neither the crew nor I knew whether I would be mad or laugh, but I decided to laugh, to the crew's relief. With my christening out of the way, we got back to work. I have no idea if the christening tradition was made up that day or not, but in a quiet act of contrition I got extra pancakes the next morning.

From the mark-capture study in June, the estimated abundance of northern pike in George Lake greater than 12 inches in size was 9,300 fish. In later years, surplus production models assured managers the abundance of northern pike was within the lake's carrying capacity, and harvest levels were sustainable.[64]

Another northern pike study I visited in 1992 was at Harding Lake. The objective of the project leader, Don, was to map fish habitat use by describing their movement and distribution in the lake relative to vegetation type and water depth. This understanding would improve our sampling strategies.

We captured twenty-six northern pike in May when sheets of thin ice still floated on the lake. Males and females were surgically implanted with radio tags 2 inches in length. Using a scalpel, we carefully made a 2-inch incision anterior to the pelvic girdle, squeezed the radio tag into the coelomic cavity, then closed the incision with several sutures. The trailing antenna was guided with a blunt needle to a tiny incision posterior to the pelvic girdle. Northern pike are hearty and were released upon their rapid recovery. The battery life of the radio tags was one to two years, so tracking their movements was a long-term commitment.

Tracking was accomplished with a scanner-programmer attached to a receiver placed in the boat. To locate fish, we took turns motoring around the lake, listening for a signal with an antenna attached to a 10-foot mast. It was a treat to get out of the office and cruise around a beautiful lake on a sunny day, listening for "beep, beep" signals.

The radio-tagged fish stayed in vegetation along the shoreline at depths less than 10 feet until early July, when they dispersed to deeper water. Large males were the first to move into deep water. Small males followed in August, but females did not move to deep water until September. Based on this study, recommendations were made to improve mark-recapture estimates of abundance.[65]

Burbot

Burbot are a tasty, white-fleshed freshwater cod, especially good deep-fried, and popular with sport fishers. In 1989, the burbot sport harvest in Region III was 4,350 fish, 90 percent of which were taken by Tanana Valley residents. Burbot were primarily sought in winter with baited set- and hand-held lines through ice holes, although summer fishing also occurred. We had several burbot assessment projects in lake and river systems throughout the region to uncover basic information such as size and age distributions, migratory and reproductive behavior, spawning characteristics, and harvest trends.

Harvest of burbot in many Interior lakes peaked in the 1980s and then declined due to overfishing. We needed to monitor stock status to determine if regulatory restrictions had allowed the recovery of these populations. One September, I joined the staff during an eleven-day sampling event to estimate the abundance of burbot in Harding Lake. It was so cold the project leader commented that we were "as stiff as the sampling gear." A crew of three in a skiff set and retrieved sixty baited steel hoop traps measuring 10 feet long. We traded off driving the skiff, recording data, working the traps, and measuring and tagging fish, so everyone could occasionally warm up their hands. Traps were fished for forty-eight hours, so we worked a systematic pattern across the lake, keeping track of the traps to be pulled at their allotted times. Our research found an estimated population of only 535 fully recruited burbot in Harding Lake, so recovery of the population was deemed modest. Set lines to catch burbot in Harding Lake continued to be prohibited.[66]

The status of burbot in streams of the Tanana River drainage was good; annual exploitation was low relative to abundance. Through tagging and radio telemetry, biologists found that movements of burbot were extensive and frequent throughout the mainstem and tributaries, so for management purposes, the entire drainage was considered a single stock. I accompanied the project leader, Matt, on both summer and winter research trips to better understand sampling difficulties. In summer, a section of the vast Tanana River was sampled for estimates of catch-per-unit-of-effort using baited hoop traps. In winter, holes were bored through the ice, and lines were set for research on reproductive biology.

Obtaining accurate abundance estimates on such a large and elusive riverine population using traditional assessment methods was difficult. I explored using a new method for estimating the abundance of the Tanana River burbot population called catch-age analysis. In catch-age analysis, population abundance is generated with a model by combining harvest-at-age data and auxiliary data, such as fishing effort. The model looked promising for estimating trends in abundance and fishing mortality; however, it suffered from small sample sizes.[67] More data is always helpful!

Arctic grayling

Arctic grayling were the most popular indigenous game fish in Region III in the 1990s, with an average annual harvest of around 22,500 fish, or 20 percent of the region's annual sport harvest of game fish. Understanding the life history of the various exploited stocks and assessing their status was paramount to their conservation. I supervised three lake and ten river research projects on Arctic grayling in the Tanana River drainage, in addition to Arctic grayling studies on the Seward Peninsula and along the Dalton Highway. At any given time, three to four fishery biologists were engaged in an Arctic grayling research project somewhere in the region, so every year, I visited at least one of them.

My daughter, Laura, holds an Arctic grayling she caught while fly fishing in the Delta Clearwater River [68]

The Delta Clearwater River is a pretty 21-mile-long stream that pours from a spring in a silvery sheet above clear beds of bright amber and hazel pebbles. It is a summer feeding area for mixed stocks of Arctic grayling that spawn and overwinter in other rivers. My family and I sometimes visited the stream during summers to enjoy the pleasant scenery, fishing opportunities, and blueberries growing along the stream's banks. The river was such a popular fly-fishing destination that high annual harvests precipitated population declines. To meet public demands for high catch rates and an abundance of large fish, the fishery became catch and release.

In 1997, I helped capture Arctic grayling in the Delta Clearwater River to estimate their abundance and composition. Sampling was conducted over two weeks in late July, with two-person crews positioned at 2-mile intervals. We used hook and line gear with flies, spinners, and jigs to capture fish which were sampled for length and age, inspected for previous marks, tagged, and released.[69]

There was a friendly rivalry between crews to see which boat could capture the most fish, and we compared notes at the end of each day. I was paired with the project leader, Bill Ridder, when one day, an Arctic grayling he had hooked gave a mighty leap into the air and threw its hook, which flew through the air at lightning speed to lodge in my cheek! We both gaped open-mouthed at an imitation mosquito attached to a hook protruding from my face. Then Ridder said, "Here, let me get that out for you." Before I could move, he leaned over and started tugging at the hook. When that didn't work, he said, "Just let me get my pocket knife out."

Aghast at the prospect of Ridder's pocket knife near my face, I brushed him away with a quick, "No thanks!" He meant well, but I was sure his further efforts would end badly for me. Ridder had not crimped the barb of his hook, so it was stuck in there pretty good. I cut the line, so he could attach another dry fly and return to fishing while I swiveled around and continued to cast my line. We fished until lunch, and then I said it was time to motor to the boat launch so I could drive to the Delta Junction medical clinic and get the hook removed.

At the clinic, the nurse cheerfully informed me that he removed fish hooks from people all the time, and with a snip and a push, out it came. By the time I got back to the river, I was bummed I had lost two hours of fishing time, and as a result, our boat might not have high count at the end of the day. I got back to the business of fishing, although I kept my back to Ridder in case his fish threw the hook.

In the fall of 1998, concerns were raised about the loss of fish habitat and the coincidental dramatic decline in the Arctic grayling population in Piledriver Slough due to the encroachment of beaver dams. Fed by an upwelling aquifer, the stream steadily suffered from a lack of flushing flows to clear out sediment and a build-up of algae in riffles and pools. The beaver dams just made it worse. While beaver dams produce benefits for fish, such as pool habitat for juveniles, the dams block fish migration to spawning and feeding areas and contribute to the eutrophication of the stream. A little less than half of the original habitat used by Arctic grayling in 1991 remained by 1998. Managers were concerned that the construction of more dams would lower the carrying capacity for Arctic grayling in Piledriver Slough.

We brainstormed a plan to identify beaver dams for removal and study instream habitat and fish population responses. First, I contacted biologists with the Game Division (renamed Wildlife Conservation) about issuing early-season trapping permits to remove beavers. They thought it was a great idea, as did the recreational trappers, who successfully removed several beavers from the stream. Next, the primary landholder of the riparian habitat affected by beaver dams was contacted, Eielson Air Force Base, about using dynamite to blow up the dams. We were pretty excited about it, but sadly, the Air Force thought it was a bad idea.

In the end, a group of us converged on the stream for the hand-removal of two of the eight beaver dams located in the stream's lower reaches. The saying, "Busy as a beaver," was true! We were dismayed at the vast amount of woody material and dirt used to build the dams. We whittled away at the impediments using hand tools (chainsaws, axes, shovels, Pulaskis, and crowbars) and a 10,000-pound truck winch tethered to a large grappling hook. After extensive effort over several days, a 25-foot section of each dam was breached, re-claiming 1.5 miles of stream flow and re-establishing five riffle sections. Arctic grayling re-colonized the exposed stream area. Because beavers could quickly repair the breaches, a sustained trapping effort was needed to maintain our progress in restoring sections of fish habitat.

I arranged for a technician, Klaus, to study the response of the slough's fish populations to beaver dam removal and apply the study towards his Master's degree through the Alaska Cooperative Fish Research Unit. He found a notable increase in Arctic grayling in reclaimed habitats.[70] Klaus rose in the ranks to become a management supervisor with the Sport Fish Division.

Sheefish

Nicknamed the "tarpon of the north," sheefish are a large member of the whitefish family that live to thirty years and grow up to 60 pounds in weight. In September 1995, I returned to the Northwest Arctic to help on a project estimating the abundance and composition of sheefish in an 80-mile stretch of the upper Kobuk River, thought to contain spawning congregations.

Because the Kobuk population contains the largest sheefish in Alaska, they are a highly sought trophy by anglers. Estimated sport harvests in the Kobuk River from 1990 to 1994 averaged about 1,000 fish and accounted for 23 percent of the state's sport catch of sheefish annually. In addition to supporting a unique sport fishery, Kobuk River sheefish were taken in an unregulated subsistence fishery and a relatively undocumented commercial fishery in the Kotzebue District, with reported annual subsistence and commercial harvests over 30,000 and 1,200 fish, respectively. Concerns for the continuance of this notable stock of sheefish and trophy sport fishery prompted our research project.

My husband, Jim, holds a sheefish he caught while sport fishing in the Kobuk River [68]

From Kotzebue, I flew to the village of Kobuk, which was near the sampling area. At Kobuk, I met a crew who had arrived by boat and we motored to the campsite where ADFG, National Park Service, and U.S. Fish and Wildlife Service biologists would stay for several weeks. The late autumn air was rank with the smell of decaying vegetation, and the leaves of birch trees had thinned to a filigree of gold.

We collected fish by beach seine and angling. The beach seine method was hard work as we targeted groups of fish in swift water containing large rocks. Compounding the difficulty was frazil ice on shore that made for slippery footing. To seine, one end of the net was held onshore by two people while the boat was pushed into the current and then motored to shore, where another two people jumped out to pull in the ends of the net—pulling a net to shore with a heavy load of large, struggling fish against a swift current made for sore muscles at the end of the day! I much preferred the hook-

and-line method. When the hook was set, sometimes sheefish jumped clear of the water and danced on their tails, their opalescent sides flashing in the light as they tried to throw the hook.

Upon capture, fish were sampled, tagged, and released. Field experience increased our knowledge about better capture methods, times, and locations for stock assessment. In successive years of study, ADFG discovered the abundance of spawning sheefish in that 80-mile section of the Kobuk River was greater than anticipated.[71]

Tanana River Drainage Chinook Salmon

We poured time and money into the sport fishery for Chinook salmon in the Chena and Salcha Rivers, tributaries to the Tanana River. Although the Region III sport harvest of Chinook salmon is only 3 percent of the state's total annual harvest, the fisheries are popular. Because Yukon River Chinook salmon are *fully* allocated to subsistence, commercial, sport, and personal users, the stocks had sport harvest guideline ranges and spawning escapement objectives.

At that time, the sport harvest guideline ranges were 300-600 Chinook salmon in the Chena River and 300-700 in the Salcha River. We used creel surveys to monitor harvest in-season. The Chinook salmon fishery on the Chena River was spread across 40 miles with multiple access points, so the survey was costly, requiring roving technicians in boats. The fishery on the Salcha River mainly occurred downstream from the Richardson Highway Bridge that crossed the river to its confluence with the Tanana River and had two access points close together: a pull-off for shore anglers before the bridge and a boat launch across the bridge. Technicians walked back and forth across the bridge to meet anglers.

One warm, sunny day in July, I got a phone call from a member of the public who said that our creel survey on the Salcha River was causing traffic problems and she requested I do something about it. I was perplexed about the situation, so I jumped in my car and drove to the Salcha River, an hour south of Fairbanks. As I approached the bridge, I instantly saw the problem. Parked on the side of the road at the pull-off, a female creel technician had decided to spend some time sunbathing on the hood of the Fish and Game truck in a bikini top. This display was responsible for slow traffic. I explained to her that professional attire and conduct were expected when on the job, and a Fish and Game vehicle is not a sunbathing platform. No further traffic delays occurred.

Initially, Commercial Fisheries Division staff counted spawning Chinook salmon in aerial surveys. To improve accuracy in escapement numbers, Sport Fish Division biologists used mark-recapture methods to estimate the abundance and composition of spawning fish. Salmon were captured on their spawning grounds using a riverboat equipped with electrofishing gear. Hefting a Chinook salmon out of the river with a long-handled dip net was a good workout by the end of the day. Captured fish were sampled, tagged, and released. The recovery event occurred a few weeks later. The salmon had spawned and died by then, so we speared and inspected carcasses for tags.

One year, the recovery of tagged salmon in the Salcha River was confounded by a large group of half-dead fish languishing in a deep pool under swift water. Attempts to spear them were unsuccessful. One of the guys, Jerry Hallberg, had a great idea—let's use a drift gillnet to scoop them out. A group of us motored to the spot, set up camp, and deployed the rather long gillnet above the pool. I helped to hold the upriver end while the net was maneuvered around the pool with the boat and then pulled ashore.

The current's pull on that large net containing heavy salmon was incredible! I dug my heels into the gravel and leaned my taunt body at a 45° angle away from the net, pulling and holding with all my might as the great force of the river slowly dragged me down the bank. The netting venture captured salmon, so we repeated it throughout the day. That night, I tried to arrange the rocks under my sleeping bag into a flat configuration but failed, gave up, and laid down. My strained muscles protested their lumpy rock mattress by cramping. I lay there in silent agony for a while before my muscles released, and I drifted into an uncomfortable sleep. At age 40+, I wondered, "Am I getting too old for this?"

After six years of mark-recapture abundance estimates we looked at the data in chagrin: escapement objectives of 6,300 fish in the Chena River and 7,100 fish in the Salcha River had been met only half the time! The problem was that mark-recapture estimates occurred after salmon had passed through the fisheries, too late for managers to take corrective actions. To provide managers with in-season information about Chinook salmon spawning abundance, in 1993, a new project, daily tower counts, was launched.

Salmon were counted as they passed under the Moose Creek Dam on the Chena River and the Richardson Highway Bridge on the Salcha River. These sites were located above sport fisheries but below spawning grounds, so they provided timely escapement information. Light-colored fabric panels were anchored on the river bottom downstream from the towers to improve the visibility of migrating fish. Lights were suspended from the towers and used during low light. Technicians were hired for eight-hour shifts, counting salmon twenty minutes every hour to alleviate eye strain and fatigue. Daily counts were expanded through calculations to get a total count of salmon passing the tower over twenty-four hours.

I took a few of the first shifts to ensure things worked as expected. I discovered that silt drifted onto the panels, so they needed to be cleaned periodically. Leaning over a counting tower for eight hours can be tedious and sleep-inducing, so we developed ways to keep technicians alert, like providing food, beverages, and music. I had heard one story of a technician who had fallen asleep and toppled off a counting tower in Bristol Bay—I didn't want that to happen!

A drawback of a salmon counting tower is the possibility of floods. The Chena and Salcha were run-off rivers, so rainfall affected their flow volume. Sure enough, high water periodically prevented counting and resulted in a minimum escapement number in those years.

By the late 1990s, we had assembled a long data series of mark-recapture and tower estimates.[72] With the knowledge gained from advances in modeling combined with the data series, biologists proposed using a different means for managing Chinook salmon in the Chena and Salcha Rivers—a biological escapement goal, which is the number of salmon that need to spawn to provide for maximum sustained yield. ADFG determines the escapement goal from biological data, such as stock productivity.[73]

Additional salmon research projects in the Tanana River drainage were led by Lisa, a well-qualified biologist I was happy to bring onto the research team. Her projects included Chinook salmon in the Chatanika River, chum salmon in the Salcha River, and coho salmon returning to the Delta Clearwater River. These rivers supported increasingly popular salmon sport fisheries.

Unalakleet Chinook Salmon

In 1997, staff with the Commercial Fisheries Division asked us to help them investigate a Chinook salmon population in the Unalakleet River. Biologists sometimes lent each other a hand in fieldwork, regardless of their division, as an efficient way to address similar objectives in the vast Alaskan wilderness. The Unalakleet River flows from the Seward Peninsula's Nulato Hills into Norton

Sound and supports commercial, subsistence, and sport fisheries for Chinook salmon. Little was known about this population, so there were no guideline harvest ranges or escapement objectives. Commercial Fisheries Division biologists relied on catch-per-unit-of-effort in the commercial fishery for trends in abundance because murky water stymied aerial surveys, and their tower counting project on a clear tributary, the North Fork, provided an unknown proportional count.

Using radiotelemetry, we launched a two-year research project to estimate the proportion of the escapement to the North Fork tributary and main stem Unalakleet River. Knowing the proportion of Chinook salmon in the two spawning areas allows the calculation of total abundance by proportionally expanding the tower count.

In June 1997, I traveled to the Unalakleet River to work with the project leader, Klaus. The plan was to implant 150 radio tags in Chinook salmon in the mainstem river 3 miles upstream from the mouth. The area was above the in-river subsistence fishery and below the confluence with the North Fork.

The first item of business was to set up a remote tracking station to record the movements of radio-tagged fish. The station included a marine battery, a computer, a radio signal receiver, and two antennas attached to 10-foot-high poles. We aimed one antenna up the North Fork and the other up the mainstem river. The receiver was programmed to cycle through 150 radio frequencies every one and a half minutes. The short cycle minimized the chance a radio-tagged fish could swim past the tracking station without being detected. When a radio tag was identified, the computer recorded the date, time, signal frequency, pulse pattern, and antenna, indicating which fork the fish had moved up. The distribution of radio-tagged fish was further refined by aerial and boat tracking.

The next step was to capture fish. After several days of experimentation, a 170-foot gillnet set in a particular configuration was found to work. The net was fished at high tide when catch rates were highest. When a fish hit the net, a two-man crew in a boat retrieved the fish and placed it in a holding tub, where it was sampled and tagged. Capture effort was distributed proportionally to the run, which lasted through mid-July.

To radio tag a Chinook salmon, the fish was gently held in a submerged cradle while I quickly opened its mouth and slid a plastic tube through the esophagus and into the upper stomach. I then pushed the cylindrical 2-inch-long radio tag with its antenna through the tube. This procedure is called an esophageal implant and is used rather than a surgical implant because salmon do not eat on their spawning migrations. The entire process lasted one to two minutes. Salmon were released in a quiet section of the river for recovery. Since the radio tag battery lasted up to four months, there was plenty of time to track fish before they died on their spawning grounds. When feasible, radio tags were recovered from carcasses for reuse because they were expensive, costing around $200.

The project went remarkably smoothly during both years. Based on the distribution of radio-tagged fish, the proportion of the Chinook salmon escapement migrating up the North Fork was relatively consistent between 1997 (37 percent) and 1998 (40 percent). Using the North Fork tower counts, the estimated escapement for the Unalakleet River was calculated to be 11,200 fish in 1997 and 5,200 fish in 1998. Although escapement differed between years, the proportions of fish migrating up the North Fork were similar, suggesting tower counts can be expanded to give a reasonable approximation of Chinook salmon escapement.[74]

Anadromous Dolly Varden

Dolly Varden are colorful fish with pink spots, named after a girl in the Charles Dickens novel, "Barnaby Rudge," who had rosy cheeks. During Alaska's Territorial days, a bounty was placed on Dolly Varden, maligned as a voracious predator of salmon. The bounty ceased in 1941 when it was discovered that, of the fish tails turned in, most were coho salmon and rainbow trout, not Dolly Varden.

Dolly Varden have both resident and anadromous forms. While the sport harvest of anadromous Dolly Varden was small, averaging 1,300 fish annually in Region III, the fishery offered high-quality angling for large-sized fish. The state record fish from northwestern waters weighed 27 pounds. Most anglers targeted these fish on their migration from the sea to overwintering areas in rivers or on their spring migration out to sea for feeding. Because they comprise a vital subsistence food to residents of northern Alaska with a harvest of unknown magnitude, and are incidentally taken in commercial salmon fisheries, the status of populations was uncertain.

Research by the project leader, Fred, on life history, stock identification, movement, abundance, and composition over a vast geographic area took time, but details slowly emerged. For example, tagging studies showed that anadromous Dolly Varden overwintering in the Wulik River, north of Kotzebue, consists of a mix of stocks that spawn in other rivers such as the Noatak, Kivalina, and Kobuk. Tag recoveries from overwintering fish in the Wulik River were recovered in the Anadyr River in Russia, indicating spawning stocks throughout the Arctic mix. In the winter of 1988, a mark-recapture experiment found an estimated 76,900 anadromous Dolly Varden (greater than 16 inches in fork length) overwintering in the Wulik River.[75]

Because the Wulik River is one of the most important habitats for anadromous Dolly Varden in Alaska, we were alarmed in late

1989 when word came that contaminates from Red Dog, a large zinc mine operation located near the headwaters of the Wulik River, were seeping into tributaries. The acidic metal-laden waters emerging from the ore body became a major source of heavy metals contamination in the Wulik River drainage. Discolored plumes of contaminants were observed flowing into the mainstem—prime overwintering habitat. Concentrations of zinc were extremely high. The construction of a water bypass system in 1991 successfully staunched the flow of contaminants.

There was great concern about the effects of contaminants on the fish population's short- and long-term health, as well as questions about the safety of the fish for consumption. Over time, contaminants precipitated out of the water column, and the anadromous Dolly Varden population in the Wulik River continued to be a vital resource to subsistence and sport users. Because unforeseen conditions might again develop at Red Dog to threaten water quality for fish and humans, a monitoring program was installed in coordination with Habitat Division biologists.

Lake Trout

Lake trout are fragile. Populations can be easily overharvested because they are a long-lived, slow-growing, and late-maturing species that may not spawn every year. In the 1990s, the annual harvest of lake trout in the Tanana River drainage averaged 5,000 fish. At that time, we knew little about their sustainable harvest rate in various Interior lakes. The project leader for lake trout, John, embarked on a multiple-year investigation of population characteristics.[76] As details surfaced about this species, it became apparent that maximum sustainable harvest rates were being exceeded for lake trout populations in the Tanana Valley, and restrictive regulations were imposed.

Lake trout do not recover quickly from capture and handling. During investigations, capture mortality of lake trout seemed high. Any method of capture, whether it is baited traps, gillnets, hook and line, seines, or electrofishing, has the potential for negative impacts on fish, so biologists try to find the method that is the least detrimental. The apparent fragility of lake trout placed me in a dilemma. On the one hand, biologists need to assess the sustainable rate of harvest for a population to avoid overexploitation, but on the other hand, the level of capture mortality should be appropriate. The question was: What is an unacceptable level of capture mortality? For guidance, I searched the scientific literature without success. I ended up using our own research as a guide.

In a study published in 1990 about the effects of electrofishing on fish in Alaska, Sport Fish Division biologists found a rate of 13.9 percent mortality in rainbow trout and concluded this was too high. So, electrofishing was discontinued as a capture method for rainbow trout. In contrast, a rate of 3.5 percent mortality in Arctic grayling was deemed low and "acceptable."[77] Okay, I had a range. Somewhere between 3.5 and 13.9 percent, my colleagues had decided a line was crossed between acceptable and inappropriate capture mortality. After discussion with staff, I developed a policy and distributed it in a memo to all Region III biologists. Only twice in my twelve years as a research supervisor did I need to direct a halt to sampling due to excessive capture mortality of fish.

Stocking

Stocking fish to divert harvest pressure on wild stocks, provide fishing opportunities, and diversify the catch began in the Interior of Alaska in the 1950s when lakes near population centers were stocked with rainbow trout and salmon. As fishery enhancement grew in popularity and received more emphasis, the ability of hatcheries to produce fish steadily increased, fueling an increase in angler effort and harvest in stocked waters. In 1989, fish stocked in four large and over eighty small landlocked lakes and one stream of the Tanana Valley totaled 2.75 million at the cost of about $430,000 to the state. The stocked fish were primarily Arctic grayling (38 percent), rainbow trout (28 percent), and coho salmon (25 percent). Additional species were stocked on a more limited basis. The stocked fish program in 1989 generated an estimated 76,000 angler days and a harvest of 93,200 fish. At that time, the stocking program accounted for nearly half the sport fishing effort and three-quarters of the game fish harvest in the Tanana Valley.

Fish produced for the Region III stocking program during the 1990s came from a state hatchery at Clear Air Force Station, located 80 miles south of Fairbanks. Supplemental fish came from a state hatchery located at Fort Richardson in Anchorage. The hatcheries evolved from a small program developed by the military to produce rainbow trout for soldiers on base. When the state of Alaska took over hatchery operations, they recovered waste heat from military power plants to economically heat water to optimal fish growth temperature (56°F) and increased rearing capacity. Most fish produced were fingerling and sub-catchable sizes. Although they cost more to make, fish stocked at larger sizes have a higher survival rate than those stocked as fry.

The project leader, Cal, conducted annual evaluations of stocked waters by estimating the abundance, growth rate, survival, and age and size structure of the stocked fish populations, as well as recording limnological conditions. Coupled with data on angler effort and harvest, the information was used to modify the program to improve benefits while decreasing costs. In addition to routine evaluations, research experiments were carried out.[78]

Initially, stocking evaluations and experiments were under the direction of the research supervisor. However, the project leader at the time, although an excellent biologist, was skilled at evading supervision. I had worked with him in prior years and thought he would be more amenable toward a boss who hadn't once been a colleague. I enticed the management supervisor to assume responsibility for him. The shift in supervision proved to be a sound decision, and I kept my hand in the stocking experiments. Everyone was happy.

In later years, the Clear and Fort Richardson hatcheries lost their supply of warm water due to the closure of the military power plants, so the state built new hatcheries in Anchorage and Fairbanks. Since coming online in 2011, the new state hatcheries have supplied fish for stocked waters in Region III.

A Stocking Request from the Commissioner

In 1993, persuasive anglers and business entities found sympathetic allies holding political office and approached the Office of the Governor and the Commissioner of ADFG with a request to create a rainbow trout fishery in flowing waters near Fairbanks that would be self-sustaining. Rainbow trout are a highly sought game fish and attract a significant amount of angling effort, so there was a belief that the establishment of such a fishery would contribute substantially to the economy of the Fairbanks area. The Fairbanks Chamber of Commerce projected that if a rainbow trout fishery could hold 10 percent of the tourist population one extra day, it would generate $2.6 million in revenue.

Rainbow trout are not indigenous to Interior Alaska; their geographic range does not extend to 65°N. latitude. The farthest-north native population of rainbow trout in Alaska was located in the Gulkana River at 63°N. latitude. The region had stocked rainbows in Piledriver Slough and found the overwinter survival rate only 2 percent. The absence of age 0 (young-of-the-year) fish further indicated a lack of rainbow trout reproduction. Nonetheless, public interest was riveted on a riverine rainbow trout fishery, and the Commissioner wanted Region III to examine the possibility. I was given the assignment.

I had grave concerns about injecting a reproducing, nonnative species into rivers that were connected to the vast Yukon River drainage. There would be no possibility of containing adverse effects on the natural system! I set my angst aside and prepared a research plan. For two days, a team of six biologists and a biometrician debated the criteria for stream selection and the set of critical physical and biological parameters needed to establish a rainbow trout fishery. We judged overwinter survival to be more limiting than spawning potential.

We selected four streams with reasonable angler access and sufficient size to support a rainbow trout fishery for the study. Piledriver Slough was designated a validation site since we already knew the outcome of stocking rainbows there. During the winter of 1993–1994, I conducted fieldwork at these five streams to measure dissolved oxygen, mid-water and intra-gravel temperatures, and low flow volume as a percentage of average annual flow. To understand the potential migratory patterns rainbows might use to survive winter, I aerial-tracked the migration of radio-tagged Arctic grayling (as surrogates for rainbows) from each stream throughout the winter. A multi-criterion decision model was used to synthesize the data and rank streams based on potential suitability for supporting a self-sustaining, introduced rainbow trout population.[79]

Of the five streams, the Chena River ranked highest. The answer to the Commissioner's question was, "Yes, there is likely suitable habitat, but there is no evidence the population would survive and grow to the extent that could support a fishery large enough to meet public demand." I concluded my report by arguing that conservation, jurisdictional, and cost/benefit concerns were hefty and the project should not move forward. I heard no word from the Commissioner's Office for a while, and then a new governor was elected, who appointed a new commissioner, and the whole brouhaha died away.

Rehabilitation

The Chena River once supported the largest Arctic grayling fishery in North America. In 1980, anglers harvested an estimated 35,100 fish. After that, the population declined. Managers imposed regulatory restrictions to decrease harvest, and they expected the Arctic grayling population to rebound. However, by 1990, conservation concerns remained, so a zero fish bag limit was imposed. In hopes of hastening the return of the Arctic grayling population to its previous world-class status, fishery managers wanted to start a rehabilitation project. The objective of the three-year project was to supplement natural production with releases of hatchery-reared progeny from the wild stock.

In May 1992, the project leader, Bob, led a team to capture mature Arctic grayling in the Chena River and hold them in net pens until they were ready for spawning. When the fish were ripe, the eggs were fertilized, placed in containers with iced water, and transported to Clear Hatchery. At the hatchery, eggs were disinfected and placed in incubators. Upon hatching, the alevins were kept at the hatchery until they were fry, 1.5 inches in length. Of the approximately 206,000 eggs taken, only 53 percent survived to the fry stage—a discouraging start to the project.

Before the hatchery-reared fry were released into the Chena River, they were marked. The proportion of marked progeny captured compared to wild fish of the same age would indicate their survival rate. The success of the rehabilitation project depended on a thriving cohort of hatchery-reared fish.

During stock assessment research, we found only an 8 percent survival rate in hatchery-reared progeny after their first year in the river. What had caused such a low survival rate? We noticed these fish didn't grow much and were in poor condition. Additionally, they had been carried downstream quite a distance. Both observations suggested starvation and disorientation as factors contributing to their low survival.

Considering that most hatchery-reared progeny died during their first year in the river, the attempt to rehabilitate our way out of a declining wild population did not look cost-effective. Luckily, the estimated abundance of the wild population stabilized. In July 1994, the estimated abundance of Arctic grayling in the Chena River was 44,375 fish. Anglers got used to the catch-and-release fishery, and the Chena River became popular again, with angler days exceeding levels seen in the 1970s.[80]

Fishery managers presume to control the harvestable surplus through their policy choices. But, in reality, fisheries science struggles to understand the interplay between a complex aquatic ecosystem, strategic human behavior, and unforeseen changes affecting both forces. Most of the time, sound stock assessment studies provide the manager with enough information to make the right calls, but Nature and people can subvert the intended outcome with unpleasant surprises. Thankfully, lessons learned from past mistakes, advances in prediction modeling, and increasing public demand for sustainable policies are improving fisheries stock assessment and management.

New Ways to Solve Old Problems

"We are continually faced with great opportunities which are brilliantly disguised as unsolvable problems."
~Margaret Mead

In my early days as the research supervisor, I sat in meetings with managers and senior biologists and watched them make decisions. They worked with unknown or uncertain information, and because the sources of uncertainty will never be completely removed, they often made intuitive judgments about varying sources of information and had personal beliefs about the accuracy of estimates. I often heard them say, "My gut feeling is…." That is, they used their intuition and beliefs in decision-making.

I could never tell to what extent a decision was based on analysis or intuition. One problem with using intuition was that, down the road, those decisions were harder to justify than decisions based on rational analysis. Additionally, intuition depends on unique past experiences, so people can process identical observations differently and arrive at differing thoughts about them. On the other hand, everyone routinely makes judgments based on their intuition, and they trust their spontaneous feelings. I thought, "There's got to be a better way to make decisions!" and discovered there is indeed a better way—by making intuitive judgments explicit in mathematical decision models. This field of study is called decision analysis. Combining rational and intuitive decision-making into one method gives a clearer representation of the problem, thus making groups more effective in achieving desired outcomes from their decisions.

Fishery managers facing complex problems with incomplete information must choose among an array of risks. The risks include reduced stock abundance resulting in diminished economic or social benefits for society. As society has become more complex, so have the tradeoffs between risks to the fish stock and risks of diminishing benefits to fishers.

Thinking that I could use decision analysis to help fishery managers improve their chances of finding solutions to complex problems, for my Ph.D. research, I decided to apply decision analysis at the policy and field levels. I remember taking my daughter to committee meetings as a toddler and seating her at the table with a coloring book that kept her occupied while we discussed my progress.

The committee wanted me to gain experience outside of Alaska to broaden my perspective. I applied to work with the planning officer at the National Marine Fisheries Service Southwest Fisheries Science Center in La Jolla, California, a facility renowned for its management of complex fisheries. The planner, Dave, kindly accepted my proposal. I took leave from ADFG in January 1992 to work with him.

The Southwest Fisheries Science Center sits on the edge of a cliff overlooking the sunny Pacific Ocean. During lunch, staff grabbed their surfboards to ride the waves. It was an amazing facility. They were on the leading edge of innovation in complex problem-solving. Dave explained their planning system, which used mathematical models aided by computer technology in a room designed to facilitate planning. He demonstrated the planning process, explained the psychology behind building consensus, and provided material to read and watch. I talked with other staff working in nonmarket valuation, biometrics, and stock dynamics to give me further insight into research techniques. The learning experience was incredible. It was the most intense dump of information into my brain I'd ever had, and I repeatedly used the knowledge gained at the Southwest Fisheries Science Center for the rest of my career.

I was guided in tackling a complex fishery problem at the policy level by a committee member, Keith, who was faculty in the School of Management. I tested the feasibility of using a decision analysis method, the Analytic Hierarchy Process, on a contentious problem: management of the Chinook salmon fishery in the Kenai River. In addition to disagreements about the allocation of salmon among user groups, there were strongly held but differing opinions about stock assessment, regulation, enforcement, and habitat use. In the early 1990s, few decision models had been applied to fisheries and none in Alaska, so we were breaking new ground. Stakeholders in various competing interest groups agreed to work with me in identifying and prioritizing issues and options.

The Analytic Hierarchy Process identified options that had broad support.[81] The set of options commonly favored by all stakeholders is the beginning point for building consensus. In particular, most stakeholders wanted more enforcement of existing regulations, reduced property taxes for landowners who dedicated their river frontage to conservation, and increased accuracy of preseason forecasts of salmon run strength. Adopting these and similar options by regulators would attract considerable support from all stakeholders. I was excited about the results because they could be used to resolve specific issues, thus whittling away at a solution to the overall management problem.

I wanted to present a paper on our findings at a scientific conference, but the director of the Sport Fish Division at the time was suspicious of the model and nervous about upsetting the Kenai River status quo. He forbade me from presenting the paper! I was distressed at this news and appealed to my major professor, Jim, for help. Jim called the director and succeeded in calming his fears a little. The director agreed to allow me to present the paper as an affiliate of the university, not as an employee of ADFG. The Kenai River evaluation was the first of many times I applied the Analytic Hierarchy Process to address complex fishery problems in Alaska.

My guide in using decision analysis at the field level was another member of my committee, Dr. Terry Quinn, a faculty member of the School of Fisheries and Ocean Sciences. We developed a model to assess the stock status of humpback whitefish in the Chatanika River and tested the feasibility of incorporating the manager's intuition about the fish stock. Adding the manager's intuitions as a weighting system improved the model.[82] I was finally able to convert the old saying, "My gut feeling is..." into useful measurements. When our findings were ready to publish, there was a new director, and I found no resistance to listing my affiliation with ADFG.

I completed my Ph.D. in 1995. It had been a long hard road. Did my position at ADFG require a Ph.D. degree? No. The man who was hired to replace me upon my retirement had a Bachelor's degree, and he did a fine job. However, the Ph.D. gave me the extra credibility needed to navigate as a woman in a male-dominated profession.

I remember attending my first meeting of supervisory biologists in 1990. The director called supervisory biologists of Regions I, II, and III to meetings at headquarters in Juneau to discuss thorny issues. I was the only woman at the table. During meetings, I would contribute an idea to the conversation, which was ignored, and then a little while later, someone else would mention the same idea and receive kudos from the other guys. This went on a few times until I realized no one was going to speak up for me except me. As my knowledge and expertise from my studies increased, I was able to contribute increasingly innovative and technical advice to the conversations and so was granted some credence.

The year of 1995 brought surprising career diversions. I was approached by Al Tyler, Assistant Dean at the School of Fisheries and Ocean Sciences, University of Alaska Fairbanks, with an invitation to join the faculty as an Affiliate Associate Professor. A faculty member had retired and they needed someone to teach Fishery Management 401. In accepting the faculty appointment, I found teaching at night, after my workday at ADFG, facilitated a two-way path—I recruited a cadre of bright, young students into ADFG and encouraged staff to seek higher education. I remember attending a Master's degree defense of one young man who had worked for me as a technician. He gave an admirable presentation of his research. I was surprised when he sent me flowers with this note, "Thank you for pushing me to go to school." It made me feel happy to know I positively influenced his resolve to further his career. Tim eventually became a research supervisor with ADFG.

I held a special academic rank in 1998 when the Dean appointed me to the first "Meek Visiting Professor" chair. Frank Meek, a gentleman who made his money in the early years of the Kodiak king crab fishery, had endowed this position. A group of us from the university had lunch with Mr. Meek, and we listened to his tales of the glorious, old crabbing days.

Later that year, I was asked by Lou Carufel, a fisheries biologist active in the American Fisheries Society, Alaska Chapter, to run for president of the Alaska Chapter. I agreed, not thinking much would come of it, but I won the election and took the helm during 1996–1997. My fishery acquaintances greatly expanded and I was exposed to a diversity of intriguing fisheries issues, but planning the annual meeting was a ton of work! Over 100 fishery biologists came to Fairbanks in the dead of winter to hear guest speakers from Canada,

Russia, and the Lower 48 and to listen to the most recent research on various topics.

One more proposition surfaced at this time—to serve as President of the Parent Teacher Association at my daughter's elementary school. How could I pass up this opportunity to get involved in her school activities? It was fun work and a welcome change from fish!

As I entered the fourth year of my job as the research supervisor, I realized that information vital to evaluating the effects of policies on angler benefits was lacking. In Alaska, public opinion is important in shaping fisheries management policy. We regularly asked anglers about their attitude and preferences regarding fishing conditions, opportunities, and regulations through creel and postal surveys.

However, we didn't have measures of public benefits, specifically, what a fishing trip was worth in dollar terms. And we didn't know what motivated a change in fishing trips. These measures would allow us to determine if public benefits outweigh management costs and which policies optimize benefits from sport fishing. During my studies, I took courses taught at the School of Management, not the typical fare of biology or fisheries majors. These courses gave me new ideas, such as using contingent valuation and contingent behavior methods in fisheries research.

The contingent valuation method uses surveys to determine the net economic value (how much a person is willing to pay above their actual cost) anglers would place on fishing trips. For example, if the travel cost to go fishing at a particular site increased by $20, would they still go fishing? How about $30 more? The net economic value is used to examine program cost-effectiveness. The contingent behavior method asks people to predict how their behavior would change given a change in the attributes of, say, a fishing trip. For example, if ADFG stocked fewer but larger Arctic char in Harding Lake, would they fish at Harding Lake more often? Fishing trips gained or lost as a result of a management change can be used to find the best policy.

In 1995, I proposed to spearhead a multi-year (1995–1998) survey-based economic and social analysis of sport fishing in Region III.[83] The first survey was of stocked waters in the Tanana Valley. The second focused on Arctic grayling. The third was on sport fishing for salmon. The fourth targeted sport fishing for burbot, northern pike, and lake trout.

The goal was to estimate the net economic value for a range of sites and species to evaluate the benefit/cost ratio for program planning. Additionally, I wanted to estimate changes in fishing trips under different management scenarios. The question was, "Can we identify optimal policy by its influence on fishing trips?"

While the contingent valuation and contingent behavior methods had been widely used by the U.S. Department of the Interior and wildlife agencies in the Lower 48 states, they had not been applied much in Alaska, never in Region III, and were unheard of by the staff. So, the staff were suspicious. I wouldn't say that, as a group, fishery biologists embrace change. Nonetheless, I got approval for the project.

The stocked waters survey was a big success. Angling for stocked fish at five major waters in the Tanana Valley was estimated to have a net economic value of $3.6 million in 1995. The benefits from stocking outweighed the costs of hatchery production and evaluations. This was good news for the stocking program! By species, the greatest net economic value was from salmon ($13.7 million), followed by Arctic grayling ($8 million) and burbot, northern pike, and lake trout combined ($4.3 million).

Surprisingly, a large portion of anglers said proposed changes in regulations would have little effect on their number of fishing trips, suggesting that of the factors that influence fishing trips, regulations play a minor role. The ability of managers to influence anglers' decisions to take fishing trips may be overshadowed by more significant factors, such as weather, availability of leisure time, and the angler's financial situation.

What did I conclude from this project? The benefits derived from sport fishing were more than our program costs of evaluating and managing the fisheries, so that was good news. Net economic values would prove useful in resource damage cases, impact statements, or allocation trade-offs. However, the ability of the Sport Fish Division to optimize benefits through regulations looked quite limited. Did managers use the socioeconomic information in their policy-making? Yes, in regional budget planning, the net economic value of fisheries was given some weight, but conservation concerns had the highest priority.

One day, questions arose at a Board of Fisheries meeting about the biological and social consequences of a proposed change in subsistence fishing regulations, and I was brought into the conversation. A public proposal had come before the board to implement rod and reel fishing as a subsistence harvest method for salmon in the lower Yukon and Kuskokwim (Y/K) Rivers because local people were already doing it anyway, so why not make it legal? Current regulations prohibited rod and reel use for subsistence because it was supposed to be a sport fish harvest tool. If the board approved rod and reel for subsistence harvest, how would this regulatory change affect fish harvest and license revenue for local residents, non-local residents, and nonresidents?

It was a big can of worms. Half of the lower Y/K residents were in favor of the proposal, and the other half were opposed to it, so it was causing controversy in local communities. In addition to disagreement, there were fears and uncertainties. Locals feared Alaskans from other communities would flood into the Y/K area to harvest their fish. Fish and Wildlife Protection officers didn't like the proposal because how could they distinguish subsistence fishers, who didn't need a license, from sport fishers, who did need a license? The Sport Fish Division was leery because sport fish license revenue might plummet.

To understand the implications of the proposal better, the board asked me to do a contingent behavior analysis. They wanted estimates of changes in fishing trips to the Y/K area, shifts in harvest magnitude, and impacts on license revenue. My research aimed to predict the behavior of licensed anglers, contingent on a change in fishing conditions—legal rod and reel subsistence. We designed, tested, and mailed out surveys, and conducted phone interviews with a sample of 1998 Alaska sport fish license holders. A contracted economist developed models and analyzed data.

The results were as follows. There was likely to be only a slight, if any, increase in non-local Alaskans traveling to Y/K waters to harvest fish because of the considerable expense and time involved in getting to the remote Y/K area. Distance and transportation costs greatly influence visitation behavior, and the Y/K area was just not that accessible.

While the vast majority of Y/K residents reported they would not take more fish if the regulation change were to go into effect, 15 percent said they would harvest more fish—an increase of nine more Chinook salmon per household. There could possibly be significant impacts on some species in some waters resulting from a minority of Y/K residents.

With regard to license fees, 44 percent of Y/K residents would not continue to buy a sport fish license. Since only 1,736 licenses were sold in the Y/K area, that's a very small portion compared to the statewide total.[84]

My research findings were made available to the board at their March 2000 meeting. Following deliberations, the board approved rod and reel as subsistence fishing gear, to take effect in July 2000, for a portion of the Kuskokwim River area with no salmon bag and possession limits, except for some restrictions on coho salmon in the Aniak River drainage.

I felt the region's forays into new research methods, such as the Analytic Hierarchy Process, contingent valuation, and contingent behavior, had greatly expanded the ability of Sport Fish biologists to understand the effects of their decisions and policies on stakeholders and ultimately to manage the resources under their care better.

More Land and More Governance

"The secret of change is to focus all of your energy, not on fighting the old, but on building the new." ~Socrates

In early 1997, the regional supervisor went on leave and designated me as the acting regional supervisor. Things were going along fine until I got a call from the director. I figured he wanted to check in and see how Region III was doing. The conversation drifted toward Region II, headquartered in Anchorage. The director bemoaned the time biologists in Region II had to spend on the Kenai River. It was true that Kenai River sport fisheries were extremely popular, economically vital, intensively managed, closely scrutinized, and targeted for allocation battles by commercial fishers. However, every region had challenges and I wasn't sympathetic.

Then, he got to the point of his phone call. Because Region II biologists were "so busy," he was transferring the Upper Copper/Upper Susitna (UC/US) management area from Region II to Region III. I was furious, not so much about the area transfer, but because he had waited to spring this move until the regional supervisor had left town, thinking I would be an easier pushover. I hoped to disabuse him of that notion.

Over several phone conversations, the director and I framed and argued our positions and negotiated conditions of the area transfer. I told him Region III would accept the area transfer only if the necessary positions, salaries, and operating funds came with it. At first, he balked at giving Region III any additional staff or funding. I was able to wrest two management positions away from Region II, along with salaries and operating funds, to ensure a smooth takeover of management functions. The area manager and assistant would be stationed in Glennallen, a 5-hour drive south of Fairbanks.

I was unsuccessful in obtaining additional research funds, which worried me. The UC/US area, 116,400 square miles in size, includes all waters in the upper 200 miles of the Copper River and the upper 180 miles of the Susitna River, accessed by four highways and numerous secondary roads. The UC/US area is close to the largest urban centers in Alaska (Anchorage and the Matanuska-Susitna Valley), so it receives angling pressure from those communities that can lead to conservation concerns. In addition to popular resident species fisheries in the UC/US area, there were twenty-five stocked waters.

Region III research biologists would be stretched thin across Region III's burgeoning 526,000 square miles and sucked into tackling a myriad of resource uncertainties in the new area, reducing the study of sport fisheries in the rest of the region. Major research issues associated with UC/US sport and personal use fisheries at the time of transfer included the lack of total return estimates for Chinook salmon in the Copper River, depressed burbot stocks in many of the larger lakes due to overfishing, and unsustainable harvests for some populations of lake trout.

My team of research biologists in Fairbanks began operational plans for stock assessment of sport fisheries in the UC/US area while I adjusted budgets. A high priority was to write a strategic plan for research on Chinook salmon in the Copper River. The Board of Fisheries had stipulated that by 2002, ADFG would have developed a management plan for Copper River Chinook salmon in collaboration with the public. Before a management plan could be devised, much missing information and uncertainty needed to be resolved.

In August 1998, I facilitated meetings of biologists responsible for research and management of Copper River Chinook salmon and used the Analytic Hierarchy Process to establish research objectives, identify missing data, and rank options. Through our discussions, we discovered that by radio tagging Chinook salmon, we could address five issues: 1) identify the location and extent of Chinook salmon spawning areas; 2) estimate the proportional abundance of the spawning population; 3) evaluate the accuracy of the current assessment method (aerial surveys); 4) identify migration timing through the in-river fisheries and on the spawning grounds; and 5) evaluate the effects of management actions.[85] A radiotelemetry project was a cost-effective way to address multiple issues and was launched in 1999. A few years later, a revolutionary tool to assess Copper River Chinook salmon was developed when I arranged for a fisheries technician, James, to study for a Master's degree. He devised a catch-age model for estimating the abundance of Chinook salmon that proved useful to managers.[86] I was happy to serve on his committee. James rose through the ranks to become a research supervisor.

While we were busy incorporating the UC/US area into Region III, the regional supervisor retired and we got a new one. Upon his arrival in the Fairbanks office, this new kid on the block appraised my office and decided he liked it better than the space assigned to him. Leaning into my doorway from the hall, he said, "This office is better suited to my needs, so you'll have to move."

I calmly told him I would *not* move and resumed my work. After a few more attempts to get me to switch offices with him, he gave up. Despite this audacious beginning, he was a good boss and listened respectfully to my ideas about research.

I made a couple of trips to visit projects in the UC/US area. A day was spent sampling fish taken in the Copper River salmon dip net fishery, and I joined three projects on the Gulkana River to assess populations of Chinook salmon, Arctic grayling, and rainbow trout/steelhead.

As 1999 advanced, we were beginning to hit our stride in assimilating a new regional supervisor and a new area into the region when surprising news interjected confusion and forced us to rethink our plans. I can't say the news was unexpected.

On October 1, 1999, the federal government extended its jurisdiction over the management of subsistence fisheries within federal public lands to include adjacent navigable waters, called nexus. The initial confusion was in how far upstream or downstream from a federal public land nexus extended. Many fish species that are important to subsistence, commercial, and sport users migrate through waters under both state and federal jurisdiction. When I called federal colleagues to clarify who would resolve stock status uncertainties within "nexus" waters, I learned about the newly-established Federal Subsistence Fisheries Monitoring Program. The program was set up to provide information for managing subsistence fisheries by funding projects that address research priorities. They were accepting proposals for research—a new pot of money!

Over the next few weeks, I held a multitude of meetings with biologists to brainstorm ideas that were both federal subsistence and ADFG sport fish research priorities, and worked like a dog to write and submit nearly twenty research proposals from the Region III Sport Fish Division to the Office of Subsistence Management within the U.S. Fish and Wildlife Service. I was pleased that more than half were approved. Region III Sport Fish Division biologists were in the field from 1999 to 2001 on federally-funded projects that included studies on sheefish and Dolly Varden/Arctic char in northern Alaska and northern pike in the Yukon River drainage.

The Region III Sport Fish research budget soared to nearly one million dollars during this time. The research program was at its strongest position since statehood in terms of funding, staffing, expertise, and scope. After more than twenty-three years of service, I retired from ADFG in 2001 with the satisfaction of leaving the Region III research program in better shape than when I first took command.

Epilogue

I retired from my career with ADFG to relish the role of being a mom before my daughter left her childhood behind. There were precious few years left, and time was flying by. I had provided my team with opportunities for advanced training to ensure they would approach research with a high professionalism and adherence to the scientific method, so I was leaving the region in good hands. Young biologists with fresh ideas deserved a chance to show their worth, and by leaving my post, I made room for them to advance in their careers.

I enjoyed the varied wilderness adventures, biological discoveries, and economic security working at ADFG. I found it especially rewarding to support and encourage young people in their desire to advance in their careers. While there were some hard times, these memories have been brushed aside by a sense of satisfaction in achieving personal success and contributing to ADFG's mission. While my book has focused on sustaining fish and game for commercial, sport, and subsistence harvest, additional components of ADFG's mission include conserving and enhancing wildlife habitat, providing viewing opportunities, documenting customary and traditional uses, and supporting the public-driven regulatory process.

Significant technological and biological advancements have been made in assessment techniques since I was employed at ADFG. For example, Bendix sonar was replaced by dual frequency identification (DIDSON) sonar to achieve greater accuracy in counting sockeye salmon in the Copper River; and genetic stock identification is now used to estimate the Chinook salmon stock composition of Yukon River fisheries.

At retirement, I didn't leave involvement in Alaskan fisheries behind entirely. I returned to ADFG temporarily as a Fisheries Scientist in an advisory role. At the University of Alaska Fairbanks, I continued teaching required courses for the Fisheries undergraduate degree. And periodically, I consulted as an independent contractor with local, state, and federal agencies and conservation organizations in using decision analysis for strategic planning, conflict resolution, and fisheries policy development.

I hope this book gives readers a better understanding of the work conducted by ADFG biologists and inspires them to support fish and game conservation measures for the good of the people and the wild creatures they depend on. Finally, I wish to encourage those experiencing trials to persevere in attaining their goals.

References

1. Editors of Encyclopedia Britannica, "Sitka" www.britannica.com/place/Sitka

2. S. McPherson and S. Marshall, 1986. Contribution, Exploitation, and Migratory Timing of Chilkat and Chilkoot River Runs of Sockeye Salmon in the Lynn Canal Drift Gillnet Fishery of 1983. ADFG Technical Data Report No. 165, Juneau.

3. P. Merritt, 1979. Beaver Survey-Inventory Progress Report, 1969-1978, Pages 46-55 *In* Annual Report of Survey-Inventory Activities Part II. Furbearers. Federal Aid Project W-17-10. ADFG, Game Division, Juneau.

4. R. Henning, et al., eds. 1981. The Kotzebue Basin. Alaska Geographic Society. Vol. 8, No. 3. Anchorage, Alaska.

5. D. Klein, et al., 1999. Contrasts in Use and Perceptions of Biological Data for Caribou Management. Wildlife Society Bulletin Vol. 27 No. 2 pp. 488-498.

6. J. Menard, et al., 2020. The 2018 Annual Management Report Norton Sound, Port Clarence, and Arctic, Kotzebue Areas. ADFG Fishery Management Report No. 20-05, Anchorage, Alaska.

7. J. Burns, et al., eds. 1993. The Bowhead Whale. The Society for Marine Mammalogy Special Publication No. 2, vxxxi, 1-787. www.northslope.org/assets/images/uploads/International%20Whaling%20Commisssion%20History%20excerpt.pdf

8. R. Suydam and J. George. Subsistence Harvest of Bowhead Whales by Alaskan Eskimos, 1974 to 2003. Department of Wildlife Management, Barrow, Alaska. www.north-slope.org/assets/images/uploads/1974-2003%20Village_harv%20BRG12.pdf

9. M. Merritt and J. Raymond, 1983. Early Life History of Chum Salmon in the Noatak River and Kotzebue Sound. ADFG FRED Reports No. 1. Juneau, Alaska.

10. J. Raymond, M. Merritt and C. Skaugstad, 1984. Nearshore Fishes of Kotzebue Sound in Summer. ADFG FRED Reports No. 37. Juneau, Alaska.

11. M. Merritt, 1974. Measurement of Utilization of Bighorn Sheep Habitat in the Santa Rosa Mountains. Pages 4-17 *In* Proceedings of the 18th Annual Meeting, April 9-12, 1974, Desert Bighorn Council Transactions, Las Vegas, Nevada.

12. T. Brabets and J. Conaway, 2009. Geomorphology and River Dynamics of the Lower Copper River, Alaska. Scientific Investigations Report 2009–5257. U.S. Department of the Interior, U.S. Geological Survey
pubs.usgs.gov/sir/2009/5257/pdf/sir20095257.pdf

13. K. Roberson and R. Holder, 1993. Gulkana Hatchery Sockeye Salmon Enhancement Project Historical Data Report, 1973-1993. ADFG. Regional Information Report No. 2A93-30. Anchorage, Alaska.

14. Photo by J. Whitman, 1981. Glennallen, Alaska.

15. K. Roberson, et al., 1983. Sockeye Salmon Studies in Copper River-Prince William Sound in 1980. ADFG Technical Data Report No. 95. Juneau, Alaska.

16. M. Merritt and K. Roberson, 1986. Migratory Timing of Upper Copper River Sockeye Salmon Stocks and its Implications for the Regulation of the Commercial Fishery. North American Journal of Fisheries Management 6:216-225.

17. R. Randall, et al., 1982. Prince William Sound Area Finfish Management Report, 1981. ADFG Commercial Fisheries Division. Cordova, Alaska.

18. R. Randall, et al., 1983. Prince William Sound Area Finfish Management Report, 1982. ADFG, Commercial Fisheries Division. Cordova, Alaska.

19. L. Janson, 1975. The Copper Spike. Alaska Northwest Publishing Company, Cordova, Alaska.

20. Alaska Fisheries Sonar. Sonar Technology Tools. www.adfg.alaska.gov/index.cfm?adfg=sonar.sonartools

21. Alaska Department of Public Safety Recruit History. dps.alaska.gov/AST/Recruit/History

22. In Memoriam. ADFG. www.adfg.alaska.gov/index.cfm?adfg=about.memorial

23. M. Evenson and K. Wuttig, 2000. Inriver Abundance, Spawning Distribution, and Migratory Timing of Copper River Chinook Salmon in 1999. ADFG Fishery Data Series No. 00-32. Anchorage, Alaska.

24. M. Merritt and K. Roberson, 1984. 1983 Copper River Sockeye and Chinook Salmon Sonar Enumeration Studies. ADFG Prince William Sound Data Report No. 84-20. Glennallen, Alaska.

25. R. Randall, et al., 1984. Prince William Sound Area Finfish Management Report, 1983. ADFG Commercial Fisheries Division. Cordova, Alaska.

26. "Pilot Believed Drowned" in the Valdez Vanguard, September 28, 1983. Vol. 8. No. 38. Courtesy of A. Goldstein, Curator of Exhibitions and Collections, Valdez Museum and Historical Archive, Valdez, Alaska.

27. "Survivor of McCarthy Massacre Killed in Fatal Fire Remembered as Hero" in the *Peninsula Clarion*, January 2, 2002. peninsulaclarion.com/stories/010202/ala_010202alapm0040001.shtml#.WSePo4WcFjo]

28. Photo by N. Wolf Dudiak, 1984-85. Homer, Alaska.

29. M. Merritt, 1985. King and Tanner Crab Assessment Studies in LCI, Alaska, 1983. ADFG Informational Leaflet No. 248, Juneau, Alaska.

30. A. Davis, 1983. The Pandalid Shrimp Pot Fishery of Cook Inlet, Alaska from the Initiation of the Fishery through the Spring of 1983. ADFG Informational Leaflet No. 220. Juneau, Alaska.

31. P. Merritt, 1984. Pot Shrimp Index Fishing in the Southern District of Cook Inlet, May 7-11, 1984. ADFG LCI Data Report No. 84-3. Homer, Alaska.

32. P. Merritt, 1985. Pot Shrimp Index Fishing in the Southern District of Cook Inlet, May 13-17, 1985. ADFG LCI Data Report No. 85-6. Homer, Alaska.

33. P. Merritt, 1985. Pot Shrimp Index Fishing in the Southern District of Cook Inlet, September 3-7, 1985. ADFG LCI Data Report No. 85-10. Homer, Alaska.

34. A. Davis, 1982. The Commercial Otter Trawl Shrimp Fishery of Cook Inlet. ADFG Informational Leaflet No. 205. Juneau, Alaska.

35. P. Merritt, 1984. Trawl Shrimp Index Fishing in the Southern District of Cook Inlet, May 21-29, 1984. ADFG LCI Data Report No. 84-4. Homer, Alaska.

36. P. Merritt, 1984. Trawl Shrimp Index Fishing in the Southern District of Cook Inlet, October 15-20, 1984. ADFG LCI Data Report No. 84-8. Homer, Alaska.

37. P. Merritt, 1985. Summary of Male King and Tanner Crab Abundance from the 1985 Index Fishing in the Kamishak District. ADFG LCI Data Report No. 85-8. Homer, Alaska.

38. P. Merritt, 1985. Summary of Male King and Tanner Crab Abundance from the 1985 Index Fishing in the Southern District. ADFG LCI Data Report No. 85-9. Homer, Alaska.

39. State of Alaska Executive Branch. ADFG Women Employment Status December 31, 1984–December 31, 1985. Papers in the possession of the author.

40. D. Shields, D. Wickham and A. Kuris, 1989. *Carcinonemertes regicides*, Symbiotic Egg Predator of Alaska Red King Crab. Canadian Journal of Zoology 67: 923-930.

41. A. Kuris, et al., 1991. Infestation by Brood Symbionts and their Impact on Egg Mortality of the Red King Crab, in Alaska: Geographic and Temporal Variation. Canadian Journal of Fisheries and Aquatic Sciences 48:559-568.

42. D. Somerton and M. Merritt, 1986. Method of Adjusting Crab Catch-Per-Pot for Differences in Soak Time and its Application to Alaskan Tanner Crab Catches. North American Journal of Fisheries Management 6: 586-591.

43. A. Davis, 1981. Dungeness Crab of the LCI area, 1979-1980 Seasons. ADFG Fisheries Informational Leaflet No. 192, Juneau, Alaska.

44. M. Merritt, 1984. The LCI Dungeness Crab Fishery from 1964-1983. Pages 85-95 *In* Proceedings of the Symposium on Dungeness Crab Biology and Management. Alaska Sea Grant 85-3, April 1985. University of Alaska Fairbanks.

45. Photo by N. Dudiak, 1984. Homer, Alaska.

46. M. Merritt, D. Bernard and G. Kruse, 1988. King Crab Stock Assessment Studies in LCI in 1984 and 1985 and Calculation of Variance in the Historical Survey Mean Catch-Per-Pot. ADFG Informational Leaflet No. 265, Juneau, Alaska.

47. C. Lean and S. Merkouris, 1987. Norton Sound Area Shellfish Report to the Alaska Board of Fisheries, 1987. ADFG Division of Commercial Fisheries, Arctic-Yukon-Kuskokwim. Nome, Alaska.

48. C. Whitmore, et al., 1986. Yukon Area Salmon Report to the Alaska Board of Fisheries, AYK Region BOF Report No. 40. Division of Commercial Fisheries, ADFG, Anchorage, Alaska.

49. The Pacific Salmon Treaty. www.psc.org

50. M. Merritt, J. Wilcock and L. Brannian, 1988. Origin of Chinook Salmon in the Yukon River Fisheries, 1986. ADFG Technical Data Report No. 223. Juneau, Alaska.

51. M. Merritt, 1988. Origins of Chinook Salmon in the Yukon River Fisheries, 1987. ADFG Technical Fishery Report 88-14. Juneau, Alaska.

52. D. Cole, 1999. Fairbanks: A Gold Rush Town Beat the Odds. Epicenter Press, Fairbanks, Alaska.

53. F. DeCicco, M. Merritt and A. Bingham, 1997. Characteristics of a Lightly Exploited Population of Arctic grayling in the Sinuk River, Seward Peninsula, Alaska. *In* Proceedings of the Fish Ecology in Arctic North America Symposium. American Fisheries Society Symposium 19, AFS, Bethesda, Maryland.

54. M. Merritt and D. Fleming, 1991. Evaluations of Various Structures for Use in Age Determinations of Arctic Grayling. ADFG Fishery Manuscript 91-6, Anchorage, Alaska.

55. M. Merritt, 1989. Age and Length Studies and Harvest Surveys of Arctic Grayling on the Seward Peninsula, 1988. ADFG Fishery Data Series No. 79. Juneau, Alaska.

56. Exxon Valdez Oil Spill.
www.britannica.com/event/Exonn-Valdez-oil-spill

57. M. Merritt, A. Bingham, and N. Morton, 1990. Creel Surveys Conducted in Interior Alaska During 1989. ADFG Fishery Data Series No. 90-54, Anchorage, Alaska.

58. L. Timmons, 1990. Abundance and Length, Age, and Sex Composition of Chatanika River Humpback Whitefish and Least Cisco. ADFG Fishery Data Series No. 90-2, Anchorage, Alaska.

59. J. Reynolds and L. Harlan, 2021. Quick Start Guide to Electrofishing. A Smith-Root Pamphlet. www.smith-root.com/support/downloads/quick-start-guide-to-electrofishing

60. D. Fleming, 1991. Stock Assessment of Arctic Grayling in Piledriver Slough, 1991. ADFG Fishery Data Series No. 91-71.

61. L. Timmons, 1992. Evaluation of the Rainbow Trout Stocking Program for Piledriver Slough, 1991. ADFG Fishery Data Series No. 92-5, Anchorage, Alaska.

62. T. Taube, 1992. Injury, Survival and Growth of Rainbow Trout Captured by Electrofishing. M.S. Thesis, University of Alaska Fairbanks.

63. S. Roach, 1992. Injury, Survival, and Growth of Northern Pike Captured by Electrofishing. ADFG Fishery Manuscript No. 92-3, Anchorage, Alaska.

64. G. Pearse and A. Burkholder, 1993. Abundance and Composition of the Northern Pike Populations in Volkmar, George, T, and East Twin Lakes, 1992. ADFG Fishery Data Series No. 93-10, Anchorage, Alaska.

65. S. Roach, 1993. Movements and Distributions of Radio-Tagged Northern Pike in Harding Lake. ADFG Fishery Data Series No. 93-12, Anchorage, Alaska.

66. J. Parker, 1993. Stock Assessment and Biological Characteristics of Burbot in Fielding and Harding Lakes During 1992. ADFG Fishery Data Series No. 93-9, Anchorage, Alaska.

67. M. Evenson and M. Merritt, 1995. CPUE Estimates and Catch-Age Analysis of Burbot in the Tanana River Drainage, 1994. ADFG Fishery Data Series No. 95-37, Anchorage, Alaska.

68. Photo by Jim Lund. Fairbanks, Alaska.

69. W. Ridder, 1998. Abundance and Composition of Arctic Grayling in the Delta Clearwater River 1996 and 1997. ADFG Fishery Data Series No. 98-35, Anchorage, Alaska.

70. K. Wuttig, 2000. Influences of Beaver Dams on Arctic Grayling in Piledriver Slough, 1998-1999. ADFG Fishery Data Series No. 00-01, Anchorage, Alaska.

71. T. Taube and K. Wuttig, 1998. Abundance and Composition of Sheefish in the Kobuk River, 1997. ADFG Fishery Manuscript Report No. 98-3, Anchorage.

72. L. Stuby and M. Evenson, 1998. Salmon Studies in Interior Alaska, 1997. ADFG Fishery Data Series No. 98-11, Anchorage.

73. Biological Escapement Goal. www.adfg.alaska.gov/index.cfm?adfg=wildlifenews.view_article&articles_id=123

74. K. Wuttig, 1999. Escapement of Chinook Salmon in the Unalakleet River in 1998. ADFG Fishery Data Series No. 99-10, Anchorage.

75. A. DeCicco, 1990. Northwest Alaska Dolly Varden Study, 1989. ADFG Data Series No. 90-8, Anchorage, Alaska.

76. J. Burr, 1990. Stock Assessment and Biological Characteristics of Lake Trout Populations in Interior Alaska, 1989. ADFG Fishery Data Series No. 90-33, Anchorage, Alaska.

77. R. Holmes, et al., 1990. Electrofishing Induced Mortality and Injury to Rainbow Trout, Arctic Grayling, Humpback Whitefish, Least Cisco, and Northern Pike. ADFG Fishery Manuscript No. 90-3, Anchorage, Alaska.

78. C. Skaugstad, P. Hansen and M. Doxey, 1995. Evaluation of Stocked Game Fish in the Tanana Valley, 1994. ADFG Fishery Data Series No. 95-20.

79. M. Merritt, 1994. Ranking Selected Streams in Interior Alaska on the Basis of Suitability for Sustaining an Introduced Rainbow Trout Population. Natural Resources, Economic Development and the Environment: The Role of Management Science, The Institute of Management Science XXXII International Conference, Anchorage, Alaska, 1994.

80. R. Clark, 1995. Stock Status and Rehabilitation of Chena River Arctic Grayling During 1994. ADFG Fishery Data Series No. 95-8, Anchorage, Alaska.

81. M. Merritt and K. Criddle, 1993. Evaluation of the Analytic Hierarchy Process for Aiding Management Decisions in Recreational Fisheries: A Case Study of the Chinook Salmon Fishery in the Kenai River, Alaska *In* Proceedings of the International Symposium on Management of Exploited Fish Populations. Alaska Sea Grant College Program AK-93-02.

82. M. Merritt and T. Quinn II, 2000. Using Perceptions of Data Accuracy and Empirical Weighting of Information: Assessment of a Recreational Fish Population. Canadian Journal of Fisheries and Aquatic Sciences. 57:1459-1469

83. J. Duffield, M. Merritt, and C. Neher, 2002. Valuation and Policy in Alaskan Sport Fisheries *In* Recreational Fisheries: Ecological, Economic and Social Evaluation. Eds. T. Pitcher and C. Hollingworth. Blackwell Science, London, England.

84. J. Duffield, C. Neher and M. Merritt, 2000. Effect of Proposed Changes to Rod and Reel Subsistence Harvest Regulations in the Lower Yukon/Kuskokwim Area: Surveys and Analysis. ADFG Special Publication No. 00-02, Anchorage, Alaska.

85. M. Merritt, 2000. Strategic Plan for Chinook Salmon Research in the Copper River Drainage. ADFG Special Publication No. 00-03, Anchorage, Alaska.

86. J. Savereide, 2001. An Age-Structured Model for Assessment and Management of Copper River Chinook Salmon. M.S. Thesis. University of Alaska Fairbanks.

Hacienda Senior Care
in Hemet
$3400/mo for Hank

Continuing Care Commty
in Boise

The Terraces
5301 E Warm Springs
Depends how much his LTCI covers
wether affords
Run by a non profit
v. high ratings

Karcher Estates Retirement + Rehab

Others with both A.L. and Mem. Care
Overland Court Sr Living 4.1*

Streamside 3.7*

Ashley Manor (at difft locations)

Aspen Creek Caldwell

Aspen Valley Sr Living

Foxtail Assisted Living in Eagle

Grace A.L. and M.C. in Caldwell

Grace " " at Fairview Lakes in Meridian

Grace at State Street in Boise

Meadow View A.L. in Emmett

Meridian Meadows Senior Campus

Paramount Parks A.L.

Park Place A.L.

Prestige A.L. at Autumn Wind in Caldwell

Spring Creek Eagle Island

Spring Creek in Meridian

Spring Gardens Meridian

Swan Falls - in Kuna

The Cottages at Columbia

The Cottages of Boise
" " Emmett
" " Meridian

The Gables of Caldwell

The Pointe at Meridian

Touchmark at Meadow Lake Villg in Meridian

Truewood by Merrill

Veranda at Paramount in Meridian

Veranda Senior Living at Barber Station

Area Agency on Aging of North Idaho - in C.D.L.
8:30-4:00 M-F
208-667-3179

Made in the USA
Las Vegas, NV
21 November 2023

81271911R10164